Worshiping

with

Charles Darwin

Sermons and Essays
touching on matters of faith and science

Dr. Robert D. Cornwall

Energion Publications
Gonzalez, Florida
2013

By combining sermons, blog posts and newspaper pieces, Bob Cornwall makes an accessible and compelling case for the compatibility of religion and science. He succeeds in accomplishing his goal of demonstrating that there is an intellectually and spiritually satisfying middle-ground between the extreme positions espoused by fundamentalist Christians and fundamentalist atheists. Contrary to what many creationists seem to believe, Cornwall argues convincingly that adopting bad science does not make for good theology. His is a powerful and persuasive voice for the goals of The Clergy Letter Project and Evolution Weekend. If you ever had any doubt that religion and science could coexist, Cornwall will set your mind at ease.

Michael Zimmerman
Founder and Executive Director, The Clergy Letter Project

Bob Cornwall's sermons read like a great conversation with an articulate and well-read friend. The prose is lively! And the perspectives he offers on creation, science, Scripture, and God are greatly needed in an age of confusion over issues in science and theology. This book is an outstanding contribution to a better way!

Thomas Jay Oord
Northwest Nazarene University
Author, *The Nature of Love* and other books

Bob Cornwall's book, *Worshiping With Charles Darwin*, offers a compelling account of one Christian and clergy member's attempt to not let go of either science or faith.

We live in an era when much harm has been done to not only the public understanding of science, but also to religious traditions themselves, by proponents of misinformation about topics like evolution. It is a sheer delight to have such an accessible collection of what a well-informed member of the clergy has written and spoken on the subject, as evidence that the wedge some drive between religion and science is not only unnecessary, but easily removed, leading to a more vibrant, compelling, and meaningful worldview.

I trust that Christians interested in the intersection of religion and their own faith will find the volume not only interesting, but personally helpful and inspiring.

Dr. James F. McGrath
Clarence L. Goodwin Chair in
New Testament Language and Literature
Department of Philosophy & Religion
Butler University

At the National Center for Science Education we frequently receive inquiries from members of the clergy seeking information about evolution and climate change. Often they seek a perspective on particular issues at the interface between religious belief and the world of science, or a recommendation of resources that can be used in a homiletical or liturgical setting. Robert Cornwall's *Worshiping with Charles Darwin* is precisely the sort of book to which I can now happily direct pastors looking for such perspectives. This excellent collection of sermons and essays addresses the question of why Christians (and people of other faiths) should embrace the evolutionary perspectives. Rejecting both atheistic scientism and the god-of-the-gaps theology of the "intelligent design" movement, Cornwall cogently defends the theological perspective of a God who acts in, with, and under the cosmic and biological evolutionary processes we observe. The author also makes a powerful case for it being of crucial importance for religious communities to come to understand the human responsibility for climate change, resource depletion, and habitat destruction. I highly recommend this collection for pastors in any denomination who would like to draw their congregations into a vigorous engagement between the realms of scientific discovery and religious belief.

Peter M. J. Hess, Ph.D.
Director of Outreach to Religious Communities
National Center for Science Education, Oakland, CA

Cover Design: Carol Everheart Roper (photomd.net)

ISBN10: 1-938434-72-2
ISBN13: 978-1-938434-72-3
Library of Congress Control Number: 2013942953

Energion Publications
P. O. Box 841
Gonzalez, FL 32560

850-525-3916
energionpubs.com
pubs@energion.com

TABLE OF CONTENTS

INTRODUCTION

I believe in God, the Creator of the Heavens and the Earth. This is a confession of faith that I share with myriads of monotheists, including Christians, Jews, and Muslims. So, how can I worship with Charles Darwin? Isn't Darwin one of the arch-enemies of people of faith? Isn't he the paragon of atheism? In answer to these questions, I will admit that Charles Darwin wasn't an orthodox Christian, but there's enough evidence that suggests that even if he walked the line between atheism and faith, he never insisted that his theories regarding evolution should be taken as evidence against belief in God. So, I can envision him sitting alongside me in the pew singing hymns of praise to God or sitting listening attentively to my sermons. I can't judge Darwin's faith. He's not here for me to query on this subject, and I will leave it to others to reflect on this question.[1]

I do affirm the proposition that God is the Creator of all things, but I also affirm the proposition that evolution is the most compelling scientific explanation for how things have come to be on planet earth. I recognize that religion and science offer different explanations for why things are the way they are, but I don't see why we must insist, as some do, that one must choose between faith and the findings of contemporary science.

I haven't always felt this way. Like many I struggle to make sense of what seemed to be competing visions of reality. One seemed naturalistic and excluded God from the picture. The other vision pictured God at the center, acting decisively to create the world that we know. Even though what I had come to believe about God the Creator seemed to contradict the science I learned in school, being

---

1   On Darwin's own struggle with matters of faith see Karl Giberson's book *Saving Darwin: How to be a Christian and Believe in Evolution*, (San Francisco: Harper One, 2008).

a good evangelical Christian I chose to side with God over science. Now I didn't want to totally reject science – that would make no sense at all. So, I had to embrace a form of "science" that fit the biblical scheme. Therefore, I began to read "scientific creationist" literature. The authors of these books, such as Henry Morris and Duane Gish, seemed to have legitimate scientific credentials, and they appeared to offer compelling answers to my scientific questions. My hope was that they could help me reconcile God and science, even if the science they offered was a "minority position." What I didn't realize at the time was that they had redefined science in order to do this.

Over time, I began to see the holes in their proposals. Perhaps it was one too many Ark sightings or the falsified attempts to read human footprints into the Paluxy River dinosaur tracks that caused me to lose faith in this method of explaining the formation of the earth and the creation of the varied species on earth. I also began to realize that one needn't read Genesis 1-11 in literalistic fashion in order to be faithful to the text of Scripture. Interestingly, it was in the reading of conservative "old earth" Creationist texts that I discovered that there was more than one way of reading the biblical text.[1] Having my eyes opened to these possibilities, I discerned that one needn't reject evolution to be true to one's faith – even a biblically defined faith. Evolution can be godless, but it needn't be!

Even if I've moved away from this anti-evolution position, making peace with both the biblical story and the scientific one, polls suggest that a majority of Americans seem content to embrace young earth creationism. This might not be a problem for many, except that it is an expression of a broader anti-intellectualism that

---

1   Among the authors who helped me let go of my young earth views was Davis A. Young, a geologist teaching at Calvin College and son of a leading conservative evangelical Bible scholar, E. J. Young. See Young's book *Creation and the Flood,* (Grand Rapids: Baker Book House, 1977) and *Christianity and the Age of the Earth,* (Grand Rapids: Zondervan Publishing Company, 1982).

is afflicting our culture. Not only are people skeptical of evolution, which is seen as godless, but it has undermined the credibility of scientific claims on issues such as climate change. And the fear that science could undermine faith may be one reason why Americans seem to shy away from the sciences as an appropriate area of academic interest. The problem might not be the school system. It could possibly be an unwillingness on the part of a growing sector of Americans to embrace the study of the sciences.

Still, Americans seem to need some kind of scientific explanation, even if what they turn to is pseudo-science. Rejecting what they believe to be naturalistic methods of the scientific community that undermine faith, they turn to alleged methodologies that allow God sufficient space to act (supernaturally). I've already mentioned "scientific creationism," which seeks to explain the formation and population of the earth within the boundaries of a 6,000 year time frame, with the primal act of creation occurring over six earth days. For some time efforts have been made to insert this "explanation" into public school science curriculums.

When attempts at imposing this "scientific creationism" failed due to its overt religiosity, activists turned to a more sophisticated version called "intelligent design." To the chagrin of proponents of this idea, in a major legal defeat, a Court decided at the famed trial at Dover, Pennsylvania that "Intelligent Design" was more religious than scientific, and thus inappropriate for the public school curriculum.[1] This hasn't stopped activists from seeking to mandate the teaching of some form of creationism in the nation's science classrooms. If this fails, they attempt to remove evolution from state curriculum standards. Activists have pursued these courses of action for specifically religious reasons. It's not about science – it's about

---

1   On the debate over Intelligent Design and the classroom, see Edward Humes, *Monkey Girl: Evolution, Education, Religion, and the Battle for America's Soul*, (New York: Ecco, 2007), and Eugenie C. Scott and Glenn Branch, editors, Not *in Our Classrooms: Why Intelligent Design Is Wrong in Our Classrooms.* (Boston: Beacon Press, 2006).

specific views of God, and the belief that an unchallenged teaching of evolution will undermine what is taught from the pulpit. It was hoped that "intelligent design" was secular enough in its tone that it could pass muster, but as the judge discerned in the Dover case, there isn't enough science in this theory to warrant its inclusion in our science curriculums.

Why did I reject creation science, young earth creationism, and intelligent design? For one thing, as I've already noted, I began to discover alternative ways of reading Scripture and theology that could make sense of modern science without jettisoning God. I discovered that attempts to fit science into an ancient creation story did damage both to science and to the integrity of the biblical story. I also began to realize that there wasn't much science involved in these efforts, and that adopting bad science didn't make for good theology. Beyond these concerns I began struggle with the perception on the part of many that Christians tended to be anti-intellectual, and I found this to be repellent.

I take seriously the command to love God with heart, soul, and mind. Faith may at times be above rationality, but it needn't be contrary to reason. As I became more comfortable with open questions, I also became increasingly uncomfortable with attempts to plug gaps in our knowledge with God. As Dietrich Bonhoeffer so pointedly stated, a "God of the Gaps" theology was dangerous to the soul.

It has again brought home to me quite clearly that we shouldn't think of God as the stopgap for the incompleteness of our knowledge, because then—as is objectively inevitable — when the boundaries of knowledge are pushed ever further, God too is pushed further away and thus is ever on the retreat. [1]

---

1 Dietrich Bonhoeffer, *Letters and Papers from Prison (Dietrich Bonhoeffer Works)*, edited by John W. DeGruchy; translated by Isabel Best, Lisa E. Dahill, Reinhard Kraus, and Nancy Lukens, (Minneapolis: Fortress Press, 2009), 8:405-406.

In the quest for truth, as a person of faith I cannot, with good conscience, embrace a stopgap God who continually retreats from reality. It is true that belief in God requires us to take a leap beyond what can be empirically proven, but it needn't move into a realm where the mind is of no value.

I am not a scientist. In fact, my scientific training is rather meager. There is much about the scientific process that lies beyond my comprehension, but then again I'm not alone in this. Instead of coming to this conversation as a well-trained scientist, I enter it from the perspective of the theologian and the pastor. I am a preacher, an interpreter of Scripture and the traditions of the Christian faith. When it comes to scientific matters I must trust what I learn from those who have devoted their lives to the scientific pursuits, especially those persons who also make a claim to Christian faith. While it's true that many scientists aren't especially religious, we're blessed with figures such as John Polkinghorne and Kenneth Miller, who are both followers of Christ and distinguished scientists. It is scientists such as these two, among others, to whom I turn for expert guidance in matters of faith and science in my own quest for truth. And Polkinghorne has the added benefit of being both scientist and theologian.

Ultimately, as a non-expert, I must trust that the scientific consensus can't be completely wrong. There is always more to discover and learn, and while there are always minority positions that need to be considered, in most cases the majority view is the more likely explanation. If not, then the minority position will marshal its facts and overturn the consensus.

This debate over the relationship of faith and science isn't a new phenomenon. Long before Darwin came on the scene, Galileo had his run-ins with the church. Before that St. Augustine found it necessary to remind religious partisans that they shouldn't look to Genesis for scientific guidance. In many ways Augustine raised questions about the "God of the Gaps" long before this became a major topic of conversation.

While these debates and arguments show no signs of abating, and while the divide can seem unbridgeable, there are many of us who not only want to achieve a cease fire, but wish to build bridges across the divide. While recognizing that science and theology offer different perspectives on reality, we needn't engage in these conversations in isolation from each other. Indeed, both modes of conversation likely will benefit from some cross-pollination. Science can expand the boundaries of our theological conversations, contributing to such discussions of issues such as the environment, gender, or sexual orientation, while theology might offer ethical/moral input into the scientific conversation.

Equally important, from a faith perspective, I believe that the future of the church and the Christian faith requires that we build these bridges. An anti-intellectual faith will eventually lose its appeal. If people must choose between what the sciences appear to demonstrate and what they believe their religion teaches, then in the long run I believe science will gain the upper hand, and God will continue to retreat from view. But such a choice isn't necessary.

Many turn to God to fill the gaps in our knowledge. This may work out for a time, but every time science fills a gap, God gets pushed further back into the realm of irrelevance. Where once God was counted on to guide every aspect of our existence, once the mechanistic view of nature that emerged with the Enlightenment, God has often been relegated to the role of divine repairman – if that. In many ways God is relegated to turning on the switch, and then going back to bed. Ultimately, such a God becomes as lonely as the Maytag repairman. But is such a vision of God worth embracing? Does such a vision of God invite our worship? In fact, if God is only of use for an hour on Sunday morning to make us feel better about ourselves, then it would seem that we might just as well find better use for even that hour of the week.

There are also those persons who engage in scientific pursuits, who when faced with the prospect of choosing between their chosen profession, or what they learned in school, and the teachings of a religious entity will likely choose the former. I'm convinced that too often we've allowed those who offer this false choice, which pushes honest seekers away from the faith, a bully pulpit from which they can declare their own vision of reality. We've allowed them to propose anti-intellectual options as being the one true way of being faithful to one's religion. It's no wonder that figures like Richard Dawkins have gotten so much traction. They feel no need to consider the voices of a Polkinghorne, a Philip Clayton, or a Nancey Murphy. To the "New Atheists,' this "moderate" or "liberal" form of Christianity isn't real Christianity. As someone who lives within this moderate to liberal portion of the Christian community, I find it sad that Sam Harris and Ken Ham are in agreement on what true Christianity teaches.

The fact is, there are many people inhabiting the faith community who reject this false dichotomy. We have no beef with science and believe that science provides important and even essential insights into the way the world works. What would life be without the contributions of the very science some Christians seek to toss aside? Just consider the technological revolution that's been fueled by modern science, a revolution that makes life more efficient, productive, and livable. One need only think of our health care advancements, which have extended life spans, cured diseases, and made for better health overall. It amazes me that there are those who not only distrust the work of modern science, but seek to overthrow its interpretation of the world, and yet have no problem benefitting from its product. They would have us return to a pre-scientific age without considering the consequences.

For those of us who live in between these two poles – the ones set out by Sam Harris and Ken Ham – there is a choice. We can continue to allow these two voices room to define the conversation, or we can reclaim it. That is, we can offer a different way of looking at the world and at faith.

Feeling that this "war" between faith and science had gone on too long, and that it was proving destructive to both science and faith, I was among the early signatories of a document called the "Clergy Letter: An Open Letter Concerning Faith and Science." There is a version for Christian Clergy, as well as Jewish, Unitarian-Universalist, and Buddhists. I was also among the original participants in what has come to be known as Evolution Weekend. This annual observance began as Evolution Sunday and was observed on the Sunday closest to Charles Darwin's birthday. For religious groups to use Darwin as a focal point of its celebration may sound odd to some, but since he is often the catalyst for debate, it seems appropriate to use his birthday as a starting point. And as a result many of us have chosen to worship with Charles Darwin. We don't worship him. He's not God. But he needn't be seen as an enemy of our faith either.

I invite you to consider the words of this letter that I chose to sign, and ponder whether or not you can be a signatory as well:

> Within the community of Christian believers there are areas of dispute and disagreement, including the proper way to interpret Holy Scripture. While virtually all Christians take the Bible seriously and hold it to be authoritative in matters of faith and practice, the overwhelming majority do not read the Bible literally, as they would a science textbook. Many of the beloved stories found in the Bible – the Creation, Adam and Eve, Noah and the ark – convey timeless truths about God, human beings, and the proper relationship between Creator and creation expressed in the only form capable of transmitting these truths from generation to generation. Religious truth is of a different order from scientific truth. Its purpose is not to convey scientific information but to transform hearts.

> We the undersigned, Christian clergy from many different traditions, believe that the timeless truths of the Bible and the discoveries of modern science may comfortably coexist. We believe that the theory of evolution is a foundational scientific truth, one that has stood up to rigorous scrutiny and upon which much of human knowledge and achievement rests. To reject this truth or to treat it as "one theory among others" is

to deliberately embrace scientific ignorance and transmit such ignorance to our children. We believe that among God's good gifts are human minds capable of critical thought and that the failure to fully employ this gift is a rejection of the will of our Creator. To argue that God's loving plan of salvation for humanity precludes the full employment of the God-given faculty of reason is to attempt to limit God, an act of hubris. We urge school board members to preserve the integrity of the science curriculum by affirming the teaching of the theory of evolution as a core component of human knowledge. We ask that science remain science and that religion remain religion, two very different, but complementary, forms of truth.[1]

I signed this letter because I affirmed its message, believing that too much misinformation was being promulgated in the name of Christianity. I did so because I felt that too many people who share my faith were rejecting science in the mistaken belief that to do otherwise endangered their faith. My concern was that this belief would undermine not only the intellectual credibility of science, but of the Christian faith as well.

Over the years, as I've participated in this observance, I have focused sermons and essays on this topic. Because I'm not trained in the sciences, I don't pretend to have answers to the great scientific questions of the day. I must entrust this work to trained scientists, trusting that where there is a consensus, truth likely will be found. This requires trust, even faith – but not blind faith. I understand that what is known today may be challenged tomorrow. So, building on that trust, I've tried to bring my own theological and spiritual understandings into the conversation.

In the pages that follow you will find a number of essays and sermons, some of which were delivered specifically for Evolution Sunday. You will find that these sermons and essays overlap each other. In some cases, such as with Genesis 1, I have addressed the text on more than one occasion. I considered choosing just one

---

1 *(http://theclergyletterproject.org/Christian_Clergy/ChrClergyLtr.htm)*

sermon from among those looking at Genesis 1, but decided that the sermons were sufficiently different to include all of them.

My hope is that these sermons and essays will encourage deep consideration of the things of God, believing with the Psalmist that "the heavens are telling the glory of God; and the firmament proclaims his handiwork" (Psalm 19:1 NRSV). Because many of these sermons and essays were inspired by my participation in the observance of Evolution Sunday/Weekend, I've added an appendix that contains liturgical resources. Let us continue seeking to better understand this faith we profess, to the glory of the Living Creator.

Sermons

1

GOOD BEGINNINGS

Genesis 1:1-2:4

Parents want their children to have a good start in life. They hope for uneventful births, proper growth patterns, and overall good health. As time goes on parents eagerly await a child's first steps and first words – hoping that they won't be absent from these milestone events. When these milestones occur, parents stop to say: That's good!

In some ways this might describe God's experience of creation as described in the opening chapters of Genesis. One could say that as God looked out and viewed the chaos that was present, God might have said – that's simply not good. And so God brought order and life to the chaos, and each step along the way, God said: That's good! Indeed, this creation of mine is very good!

Genesis 1 captures this sentiment in the form of an ancient poem, or perhaps more specifically, a liturgy of thanksgiving. When we read it out loud, it should become apparent to us that this is not intended to be taken as a science lesson or as a historical account. It would be appropriate to say that this is a call to worship that calls on us to recognize the order and beauty of the universe. Rather than telling us how and when the world was created, the text suggests that this world we live in has meaning and purpose. Science, for its part, offers us the answers as to how this process worked out. You could say that while science deals with the minutiae, Scripture gives the big picture. I know that there are many Christians who look to Genesis 1 for scientific and historical answers, but in doing this I think they miss the point of the passage.

When we hear this liturgy of praise, we need to understand its context. Most creation stories of that day weren't nearly as positive as was Genesis 1. Israel's neighbors thought that the world was evil and dangerous. When you read these other stories, rarely will you hear it said: "It is good." No, what you hear is: Don't trust the gods, they're capricious and self-centered, so protect yourself. But Genesis 1 says: look around and what you see is good, and it is good because God made it good.

BEGINNINGS

Genesis is a book of beginnings. It tells us where we came from and where we are going. It is also a book of covenants, covenants between God and humanity. God establishes covenants with us because God wants to be in relationship with us. Genesis 1 tells us that we are a reflection of God's glory. We are the image of God. We may walk away from God, but from the very beginning, God chooses not to walk away from us. That is because we bear God's image.

We are bearers of God's image and we live in a special place that God has given us. It is a place where we can grow and we can learn. It is a place for us to enjoy the presence of God. Now, Genesis doesn't tell us how this environment came into being, it just says that it came into being. Like I said earlier, I don't have any problems with the scientific explanations about the processes of an evolving creation. I'm not a scientist, but I trust its answers. Genesis reminds us that there is more to the story than what we see. It says that God gives purpose to creation. You have to grasp this truth by faith, but when you do, you will stop and give thanks.

RESPONSIBILITIES

Genesis tells the story of a covenant that God makes with creation. As the other party in the covenant we have our responsibilities. God says to us: I've given you a good beginning. I've created a world for you to live in. And it's good. Enjoy it, celebrate it, and take care of it. Use it carefully, don't abuse it. Although many people see this world as a temporary stopping point on the road to heaven, Genesis reminds us that this life is more than an interruption on our trip to heaven. And if we're going to be here for a while, we have some responsibilities of our own.

BE THANKFUL

When you go to the Grand Canyon, Waimea Canyon, Crater Lake, or Yosemite, what is your reaction? I've heard about the person who goes up to the Grand Canyon and says: Nice ditch! I don't really think it ever happens. I've not been to the Grand Canyon or Yosemite but I've been to Hawaii's Waimea Canyon and to Crater Lake. I must say – I've been favorably impressed.

When you see something of beauty and grandeur, you will stop and enjoy it. You will be filled with awe and wonder. If you're human, something inside will tug at you and you will give thanks. And that's the point of Genesis 1. Celebrate the grandeur and wonder of God's creation in all its complexity and mystery, because God created with loving care.

BE A STEWARD

There are those who say of this earth, "God gave it to us to use!" They say: you have dominion over it. You're the king; you're the queen, so do with it as you please. Nothing was meant to last forever, so use it before it's gone – the oil, the trees, the water. There are those who still question global warming, but the evidence is mounting. We're discovering that the resources aren't infinite. The world is good, but we can corrupt it. There may still be plenty of oil in the ground, but, but that doesn't mean we should use it all.

I've seen pictures of turn-of-the-20th century Summerland, California. Today it's a sleepy beach town, but back then where the beach lies today, once stood acres of oil derricks. I'd rather not see the Santa Barbara coast covered with oil wells just so people can drive Hummers.

So, did God give us unfettered dominion over the earth? Or did God call us to be stewards, to be care takers of creation? If the latter is true, then God didn't give us permission to rape and pillage the earth of its resources. Instead, God wants us to nurture, protect, and care for God's handiwork. Remember too, you and I are part of the same creative act that created the environment in which we live.

That, I think, is what God had in mind. Take care of my garden, won't you?

TAKE A REST

The first few verses of Genesis 2 say that even God rests once in a while. And if God rests, shouldn't you take a rest? In other words, stop and smell the roses.

With this in mind I have an assignment for you. Why not take a drive to Surf Beach, past the houses, the fields, down along the river, and finally to the beach. Look up and down the beach, to the river's outlet and beyond. Look out into the fathomless ocean. I think you'll be favorably impressed!

Then, at night, weather permitting, look up into the sky and examine the magnificence of what you see. Billions and billions of stars, some possibly with planets just like ours, planets that just may have some beings like us, who are doing much the same thing. Aren't you awestruck?

Finally, look around you at your neighbor; look deeply at the beauty of the human creation. You will see and you will celebrate that we are complex individuals. We have our issues, but we are here and we are evidence of the goodness of God's creation.

And God said: It is good!

Preached: May 22, 2005 (Trinity Sunday)
First Christian Church (Disciples of Christ)
Lompoc, California

2

What's With Evolution Sunday?

Genesis 1:1-5

Everybody loves dinosaurs, especially young kids. From Barney the purple T-Rex to Little Foot and his friends, dinosaurs tickle their fancy. Discovery Channel, PBS, the History Channel, all air programs that tell dinosaur stories. Then there is the Jurassic Park franchise, which was a mega-hit.

I remember my visit to Dinosaur National Monument when I was about seven. I can still remember the great wall of fossils being excavated. Brett loved the Natural History Museum at the University of Kansas. It didn't have a T-Rex, but it did have a big reptilian-like fish that swam the sea covering Kansas millions of years ago. With so much interest in prehistoric life, it seems odd that nearly half of Americans reject the science that explains the very things that fascinate us. For many people, science is either the enemy of faith or it is redefined in a way that fits a so-called biblical time-line.

Charles Darwin's *Origin of the Species* appeared a century-and-a-half ago, offering his theory of natural selection as an explanation of how things came to be. Theories of evolution have been around for centuries, but Darwin's theory revolutionized the study of human origins and development. He helped explain why there is so much diversity in earth's flora and fauna. Since Darwin's time we've continued to discover new species deep within jungles, caves and oceans.

Now, not everyone liked Darwin's solution, mainly because it seemed to conflict with the Bible. The choice was simple, Darwin or God. Of course, this wasn't the first time that science and the

Bible had come into conflict. St. Augustine turned to allegory to explain Genesis because it seemed to conflict with the science of his day. Galileo's telescope proved controversial, because his discoveries seemed to contradict the Bible. Despite the objections of some Christians, many other Christians have found ways of "adapting" to Darwin.

So, What's Evolution Sunday?

You won't find Evolution Sunday in the church's annual planning guide, but with the ongoing debate about intelligent design, creationism, and evolution, it seemed like an appropriate topic to consider. While it's not an official religious event, more than 400 congregations from around the country are observing it, and we're one of them. Evolution Sunday is an expression of the Clergy Letter Project started by a biology professor from Wisconsin. After being drawn into a debate that was wreaking havoc on the local school system, he decided it might help the cause of science to build a bridge between science and Christians. More than 10,000 clergy from across the country have signed his "Open Letter on Religion and Science," and I'm one of them. You will find a copy of this letter in your bulletin today and there is more information on the web.

We're observing this event today because I'm concerned about the growing debate over science and religion. The divide between the two communities is widening every day, and we live at a time when we need science to help solve a whole host of questions and problems. We also need the moral and ethical input of faith to help guide our use of new technology. As this debate grows louder, fewer young people are entering the sciences. I think there is a correlation between the growing rejection of evolution and these trends. There is another problem. By rejecting mainstream science, many Christians have made Christianity seem unreasonable and anti-intellectual. That's too bad, because it puts an unnecessary barrier to the hearing of the gospel.

WHAT DOES SCIENCE SAY?

I'm not a scientist, but I do read and I'm the curious sort, so I've looked into the issue. I may not understand all the details, but I get the gist of it. Science says that the universe is very old. Some say it started with a bang and others say it's always been there, only it's changed a bit over time. When it comes to humans, science says that we developed over millions of years from a very primitive ancestor. Some people don't like this. I guess they don't like having an ape in the family tree; plus it seems to conflict with Genesis.

The problem is, there is overwhelming evidence to support evolution's claims. There have been many challenges to aspects of the theory, but none have replaced it. The piece of evidence that I find most convincing is one that affects our daily lives. Evolution lets medical researchers use animals, like mice and chimps, to test medicines that save human lives. These tests wouldn't work if we didn't all share some common DNA. Darwin didn't know about DNA, but it supports his theory. So, if you like modern medicine, you've got to like evolution.

WHAT DOES THE BIBLE SAY?

How do we reconcile the facts of science with our faith and our reading of Scripture? Do we have to choose between God and science? Genesis says that in the beginning God created the heavens and the earth and it uses a six-day pattern to describe the creative acts. Finally, on the seventh day God rested. There is a constant refrain in Genesis 1 that seems important. That refrain is: "It is good." This passage suggests that human beings are created in the image of God, but what does this mean? Does it mean that we look like God? Does it mean that we share God's intellect? Or does it mean that we serve as God's icons or representatives in creation? These are good questions.

Then there is the question of how we read we read Genesis 1-2. Must we read Genesis as a scientifically and historically accurate narrative to be true to its intent? Or, should we read these creation stories as a theological statement? When I read the opening chapters of Genesis, I don't find answers to scientific questions. That's because the writers and the recipients weren't asking scientific questions. What Genesis does is challenge another creation story, the Babylonian one. In that story God isn't the creator, the creation is God. In fact, in this story, the sun, the moon, the earth, and the sea are different gods, and these gods aren't necessarily good. While the Babylonian story instills fear, Genesis calls for praise and adoration.

So, here we are on Evolution Sunday. We have a choice, we can build a wall between faith and science, or we can build a bridge. I think building a bridge would be more productive, because it will lead us to giving praise to God. In the mean time, let us rejoice because the "heavens are telling the glory of God; and the firmament proclaims his handiwork" (Ps. 19:1).

Preached: February 12, 2006 (Evolution Sunday)
First Christian Church (Disciples of Christ)
Lompoc, California

3

THE WISDOM OF CREATION

Proverbs 8:22-31

When you see the Grand Canyon, a Hawaiian sunset, or Crater Lake for the first time, it's unlikely that the first thing that pops into your mind is a science question. You'll probably say something like, "Wow, isn't that beautiful!" And then, after taking in the sights for a bit and likely taking a few pictures, you may start asking questions about the science that explains these expressions of nature's beauty. The first reaction is aesthetic, even spiritual; the second might be scientific. Science deals with the how and faith deals with the wow!

Those two different reactions suggest that there's more than one way to look at things. Neither one is right or wrong. They're just different. The scientific angle is extremely important, and it should be honored. But it doesn't always tell the whole story, for the rest of the story many of us turn to faith. In my mind, faith and science aren't competitors, they're complements.

For centuries now people have been arguing about the relationship of religion and science. Some people believe they're opposed to each other, which means you can't believe in science and be a good Christian at the same time, or you can't believe in God and be a respectable scientist. There are also people who want to merge the two, by letting religion determine what is scientific. And then there's the belief that science and religion are two different ways of looking at things. Both are valid, but they're very different.

I'm not a scientist, but I'm interested in science. I also believe in God and am very interested in matters of faith. As a person of faith who values science, I've become very concerned about a grow-

ing skepticism about science that's present in our country and find
it disheartening that much of this skepticism is rooted in religion.
I'm concerned that this skepticism can hinder important research
that would benefit our world. There are important medical issues at
stake and environmental ones as well. Solutions to these problems
will come from the scientific community. Because I believe in the
importance of knowledge and the intellectual credibility of my
faith, I've also become very concerned about a growing anti-intellectualism that's developed among many Christians.

These concerns about scientific research and intellectual credibility are why I'm preaching about Evolution Sunday once again.

If you read Genesis 1 closely, you'll discover that when God
created the heavens and the earth, God said that it was good. Genesis doesn't tell us how; it just tells us that the end product is good.
There's a scientific theory called the "anthropic principle," that
suggests that our planet is perfectly suited for human life. If things
had been just a bit different, life wouldn't be possible. And so the
question is: Is this a happy accident of some kind or is it an act of
divine providence? Without going into detail about this principle,
let me say that my belief in God the Creator, makes me comfortable
saying that this was no accident. That is of course a statement of
faith, not a statement of science. We may discover other scientific
theories that challenge the anthropic principle, but like the big
bang theory, it does offer room for conversation about God's role
in creation. As physicist and theologian John Polkinghorne says,
this suggests the possibility of "an inbuilt potentiality to creation."[1]
It is of course suggestive only and not definitive. It doesn't preclude
further study of the mechanisms of creation, but in my mind it does
cause me to stop and give thanks for God's good work.

I believe in a Creator, but as I try to understand the mechanism
of creation, I don't turn to the Bible, but to science. I have great
respect for the work of the scientist, and that's why I'm comfortable
with the theory of evolution. There is near unanimous agreement

1   John Polkinghorne, *Faith Science and Understanding*, (New Haven:
Yale University Press, 2000), p. 68

among biologists that we are a product of evolution through the process of natural selection. That's the scientific explanation and there's a lot of evidence to back it up.

But like I said, there's more than one way to look at things. Theology doesn't replace science, but it does offer a different perspective. Traditionally Christian theology has taught that God spoke everything into existence, although Genesis 2 offers a slightly different take on things. We're used to thinking in a top-down way about creation. God is something like a CEO who orders things to get done and they're done, Fed Ex style! Evolutionary theory suggests that things are a bit more complicated than that. And so I've begun to think about God not as commander-in-chief, but as persuader-in-chief.

Proverbs 8 says that the first act of creation was the creation of Lady Wisdom. Wisdom then served as God's assistant "like a master worker," whose work brings great delight to the Creator (Prov. 8:30). Proverbs 8 seems to suggest that creation is a process that takes time to develop.

In a somewhat related way, theologian Jürgen Moltmann talks about "the energies and potentialities of the Spirit" through which the Creator is present in the creation.[1] Instead of Creation being a top-down, outside-in kind of job, it's an inside-out job. Instead of God standing on the outside giving orders, God acts from within creation through the agency of Wisdom or the Spirit, seeking to persuade the very atoms and molecules to work together for the common good. Now sometimes the atoms and the molecules decide to do their own thing, but whenever they work together good things happen. And in the end the universe comes into existence. In fact, it's still coming into existence, because, God's not finished creating quite yet. Now that's a bit simplistic, but it suggests a way of thinking about creation that might be a bit different.

---

1 Jürgen Moltmann, *(God in Creation: A New Theology of Creation and the Spirit of God,* trans. by Margaret Kohl, New York, Harper and Row, 1985), pp. 9-10.

Whatever theological description we give to creation, I think that at the end of the day, what our faith does is call us to celebrate God's good gift of creation. Theology, which is rooted in Scripture, doesn't explain everything, but when we take science and theology to be complementary descriptions of reality, we can celebrate the beauty of this world as well as better understand it's complexity. There are things that I find difficult to reconcile – things like earthquakes and such. I think that Moltmann might be right that creation requires redemption and reconciliation. But that's for another day.

The writer of this proverb has it right: God does delight in Wisdom's handiwork. When we look at the big picture, we find balance and purpose to this world of ours. Science tells the story in its own way and faith in its own way. I'm amazed at what I read in the science book, but I'm also left wanting more. I want to have a conversation with the one who brought this universe into existence. And with Isaiah I want to join trees of the field in declaring God's glory. When I see a beautiful sunset, a magnificent mountain peak, or the grandest of canyons, I do want to know how this came to be, but first I want to stand in awe and give thanks to the one who is revealed in its glory. But the point of Evolution Sunday is this: Science and faith aren't enemies; they're two ways of telling the story of the universe. Because Scripture talks about meaning and purpose, it serves to remind us that this world belongs to God and not to us. We are called to be good stewards, not careless users.

Preached: February 11, 2007 (Evolution Sunday)
First Christian Church (Disciples of Christ)
Lompoc, California

4

Time to Change the Light Bulb

Genesis 1:1-2:3

God said, "let there be light and there was light." And when God saw the light, God said, "That's good!" Indeed, light is good. As you know, it's kind of difficult to see where you're going in the dark, so having some light can be helpful.

In the beginning of time, you had the sun – not bad during the daytime – and then there was the moon – it gives off some light, but it's pretty limited. In time somebody discovered fire, and fire helps a lot. From there the campfire gave way to the candle, and the candle to the lamp. Every advance in lighting made living indoors and going about at night just a bit easier. Then came the biggest revolution in lighting history. Back in the 19th century somebody figured out how to hook up lights to electricity and everything changed. Even though he wasn't the first to come up with the idea of the incandescent bulb, a guy named Thomas Edison came up with a long-lasting filament and history was made. In fact, his idea worked so well that we're essentially using the same technology today as we were in 1880. Edison's invention has been a great success, and we give thanks for it every time the lights go out and then come back on. There's just one problem; these bulbs use a lot of energy. While it's hard to give up something so successful, perhaps it's time to change the light bulbs!

There's a new kid on the block – actually a couple of new kids. Maybe you've seen them; they're called compact fluorescent bulbs (or CFL's for short) and LED's. These lights are much more energy efficient. A CFL bulb uses 70% less energy than a typical incandescent bulb. Now, I'm no math whiz, but that seems like a

big difference. Just think what would happen if we exchanged just
one million incandescent bulbs with CFL's – we could save 400,000
megawatt hours of energy in one year. That would keep 200,000
tons of green house gases from being released into the atmosphere.
The amount of energy saved could power 60,000 homes for a year,
and is the same as removing 31,000 cars from the road. And, that's
just one million light bulbs. Just think what would happen if we
all did this?

LENTEN CHALLENGE ON EVOLUTION SUNDAY

So, why am I talking about light bulbs? Besides the fact that
light is referred to in Genesis 1? The answer is that besides being
the Sunday before Valentines and the first Sunday of Lent, this
is also Evolution Weekend. Yes, we're doing Evolution Weekend
again this year, and as I was thinking about what to do with this
day, I thought – hey, let's talk about the environment. As you may
remember from last year, Evolution Weekend not only coincides
with Darwin's birthday, it is an opportunity to consider the relationship of our faith to science.

You may have heard that religion and science are at war. There
are those who say that science has made religion obsolete. There
are others, on the other side, of the spectrum, who say that Charles
Darwin, and scientists like him, is the devil incarnate. You have to
choose. And there's my problem – I don't want to choose. In fact,
I've come to believe that my faith can live in peace with science,
including evolutionary science.

Because Evolution Sunday falls this year on the First Sunday of
Lent, I thought maybe this year we could combine the two obser-
vances. Evolution Sunday challenges us to learn from the witness of
science. Yes, it challenges some of our cherished ideas, but the end
result is a stronger faith. As they say, Darwin is here to stay, so how
are we going to deal with him? Lent, on the other hand, calls us to
a time of fasting, prayer, reflection, and sacrifice. Maybe this year,
instead of giving up chocolate or Doritos, we could do something

more constructive. Perhaps we could take some time and change some light bulbs, turn off some appliances, drive a smaller car, and do something good for the environment.

OUR CONNECTION TO THE CREATION

We've already heard the grand poetic statement of Genesis 1, which celebrates God's act of creation. This passage can be taken in two ways. We can interpret it as saying – God set this earth up for you to do with as you please. You're the master and the earth is your servant. That's been a popular way of looking at things, but if we look at Genesis 1 through the eyes of the evolutionary scientist, maybe we'll begin to notice the connection between humanity and the earth. I think this connection is even clearer in Genesis 2. There in the second creation story, God takes some dust and makes a man from it. That dust of the earth is symbolic of the building blocks of the universe. The Bible and Science tell the story differently, but in each of these stories I hear the message – you are connected to the world in which you live. So be good to it.

OUR CALLING AS THE IMAGO DEI

Going back to Genesis 1, we read that God creates humanity in God's image – male and female. Bearing the image of God, which in Latin is the *imago Dei,* we're God's representatives on earth. These first humans are told to be fruitful and multiply and in most translations, they're told to take dominion over the earth. One of the reasons why our environment is under such a great threat is that too many of us take this to mean – you can do with it, whatever you want. Listening to the voice of science, I'm drawn to the idea that our calling as human beings is to take care of the creation, to be stewards rather than masters of it.

OUR RESPONSIBILITY FOR THE EARTH

We're observing Evolution Weekend again this year because God could be, and I think is, wanting to use science to tell us something. Science tells us that we're all connected. What's bad for the polar bear is ultimately bad for us. If the Arctic ice pack disappears, not only will the bears disappear, but so might cities and islands all around the world. Global Warming, Al Gore said, is an "inconvenient truth." There are those who deny that there's a problem and others say that a solution is too costly. But if we're willing to listen, science could offer us some solutions, solutions as simple as trading those old incandescent bulbs for some CFL light bulbs. If enough of us do this, then we can make a difference.

If you look at the card stock insert in your bulletin this morning, you'll find some ways of doing just that – making a difference. But I don't just have an insert for you; I actually have light bulbs to give to you. Thanks to the City of Lompoc, each of you will get to take home your own CFL bulb. All I ask is that you take that bulb and replace an incandescent one with it. Now, I know that some of you are resistant to change. You've been using those incandescent bulbs all your life and you just don't think that these newfangled bulbs will work as well. My Lenten challenge to you is to let go of those fears and make the change. At our house we've replaced almost all of the bulbs, and we've got plenty of light! So, with no further ado, and before we sing our invitation hymn, it's time to pass out the bulbs!

Preached: February 10, 2008 (Evolution Sunday)
First Christian Church
Lompoc, CA

5

HAPPY BIRTHDAY CHUCK!

Colossians 1:15-20

I want to begin this morning by giving a big Happy 200th Birthday cheer to Charles Darwin. In case you missed it, on Thursday Darwin joined Abraham Lincoln in celebrating his 200th birthday. Now neither of them was around on Thursday to share in the festivities, but we can recognize and celebrate their legacy anyway.

Now, one of my more famous predecessors as pastor at Central Woodward was a big fan of Abraham Lincoln. As I understand it, Edgar DeWitt Jones hosted an annual Lincoln Lecture, because the study of Lincoln was one of his passions. So in the spirit of my predecessor, I invite you to share in one of my passions by observing Evolution Weekend on the Sunday following Charles Darwin's birthday. This year the number of churches, synagogues, and mosques participating has grown to about 1000.

This event was born four years ago as an outreach of the Clergy Letter Project. That project produced a letter, which you will find in your bulletins this morning. The letter, which was written by Dr. Michael Zimmerman and then signed by more than 11,000 clergy and theologians, including me, is entitled "An Open Letter Concerning Religion and Science." By signing this letter we declared our belief that Christians can believe in God and also affirm the scientific truthfulness of evolution.

I realize that many Christians would disagree with that statement, and I'm sure they would find it not only odd but sacrilegious for a church to observe the birthday of Charles Darwin. After all, in the minds of many he was the spawn of Satan, and an enemy of the church. Obviously, I don't share that sentiment. It's true that

Darwin's theories have posed a challenge to our faith, and they have forced us to reconsider some of our traditional readings of the Bible, but even though Darwin was an agnostic at his death, he was never an enemy of Christianity or of the church. In fact, he remained a member of his family's church and contributed to it until his death – in honor of his wife's deep faith.

The reason why I introduced this observance to the Lompoc church and here is that I believe that something very important is at stake in this debate over the relationship of faith and science. Indeed, I believe that the intellectual integrity of our faith and our witness to the world is at stake.

JESUS, DARWIN, AND THE SPIRITUAL MIND

Although it's recorded in the gospels that Jesus told his followers to love God with their heart, soul, and *mind* (Mt. 22:37), there is a lot of anti-intellectualism present within the Christian community. Many Christians seem to be afraid of what they'll discover if they start asking too many questions about the meaning of the Bible or their own faith tradition. Better not to ask questions, and if people start asking questions, it's best to change the subject.

The reason why I'm so passionate about this issue is that I believe very strongly in the principle that "all truth is God's truth." If this is true then I believe we must, as Christians, be willing to pursue that truth no matter where it takes us, even if it takes us down paths that we find uncomfortable or challenging. The good news is that we don't have to take the journey alone. We can go on this journey together in the company of God's Spirit.

By taking this pathway, we will be true to our heritage as Disciples of Christ. The Disciples have been, from the beginning of our movement, committed to the life of the mind. Sometimes we can be overly rational, but the point is, as important as the mystical and the experiential may be to our spiritual welfare, our minds are important as well. Indeed, when we come to church, we shouldn't have to leave our brains at home!

The problem we face today as Christians is that there are too many partisans on both sides of the issue telling us that we have to choose: It's either God or Evolution. You can't have both. As for me, I reject that demand. Like many Christians, who unfortunately have been quiet of late, I want to declare my firm belief in God the Creator and at the very same time affirm the teachings of modern science concerning the manner in which this world emerged.

INTERLUDE: JESUS LOVES DARWIN

There's a bumper sticker that features two fish kissing. Maybe you've seen it. On one fish the name of Jesus appears, and on the other one, the one with legs, you'll find the word Darwin. If you go to our church Facebook page and then check out the invitation I sent out for today's service, you'll be able to see it. I used that symbol because I think it's very appropriate for what we're trying to do today.

That bumper sticker has a very ancient lineage. You see, the fish has been a Christian symbol since the first century CE. The fish reminds us that some of the earliest church leaders were once fishermen, and Jesus himself invited them to join him in fishing for humans. Of course, there's another reason they used the fish – it makes for a very nice acrostic that carries with it an important theological message: You see, the Greek word for fish is *ichthus,* and if you take each letter of that Greek word you can get this statement of faith: Jesus Christ, God's Son, Savior.

In recent years lots of fish decals have sprung up. When you see one, you expect that the person driving the car is a Christian. Because Jesus and Darwin are supposed to be at war, it's not surprising that the "other side" came up with their own similar decal. Their fish, however, has legs, reminding us that the first land animals descended from fish, and instead of Jesus' name, you'll find Darwin's name on it. By bringing these two fish together, we declare our belief that religion and science aren't enemies.

I realize I can't speak for everyone here today, but I would like to affirm this three-part premise: As followers of Jesus, who believe firmly that God is our creator, we can also affirm three important scientific premises: 1) Our universe is very old; 2) Humans share a common ancestor with all living things; and 3) natural selection is the currently accepted scientific explanation for how all of this has taken place. I realize that there's a lot more that can be said here, but I think that's a good start for now.

JESUS, CREATION, AND REDEMPTION

You might be wondering – what about that scripture text that we read today – where does it come in? That passage, the one from Colossians 1, speaks clearly and powerfully of Jesus' role in creation. It is, in fact, a hymn, a song of praise to Jesus, declaring to all that he is God's partner in the work of creation and redemption.

As to the first point, this hymn boldly declares that Jesus is the first born of creation, and that in him, and through him, and for him, all things, whether in heaven or on earth, have been created. Not only that, but he is before all things and in him all things hold together. Indeed, he is the beginning and the end, the alpha and the omega. And then the hymn moves on to the second point. In Jesus, God chose to dwell and in him and through him God has reconciled all things. That is, in and through the cross of Jesus, God has brought peace to earth and to heaven.

The language of this hymn not only soars, but it's cosmic in nature. Everything, not just our existence, is taken up into Jesus, so that everything that exists might find its purpose in God.

This passage, whether written by Paul or not, reflects in hymnic language the biblical confession that God is the creator and that what God creates is good and has purpose. At the same time, it reflects the biblical confession that brokenness has crept into this creation. Indeed, as Paul himself writes in Romans 8, the whole of creation is groaning in labor pains, anticipating the freedom and the wholeness that it will gain together with the children of God at the appointed time (Romans 8:22ff). Now that's not a scientific statement. It's poetic and theological, but nothing in that statement is at odds with science.

It's my belief that both science and theology have something important to say to us. Each bears witness to important truths, but they do so from very different perspectives. We get into trouble when we try to turn the Bible into a science book. And, while science has much to say to us as Christians, there are truths that are beyond even its insights. It doesn't make either of them deficient – just different. We can learn from both and celebrate both. And that, I believe is the point of Evolution Weekend! So, since Jesus loves Charles Darwin, we can wish him a very happy birthday!

Preached: February 15, 2009 (Evolution Sunday)
Central Woodward Christian Church (Disciples of Christ)
Troy, Michigan

6

WERE YOU THERE? – IN THE BEGINNING
Job 38:1–11

The liturgical calendar may say that today is Transfiguration Sunday, and the social calendar might say that it's Valentine's Day, but I have another calendar that says that it's Evolution Sunday. As you can see from the service, I decided to go with the latter calendar! For the fifth consecutive year churches and synagogues from across the country will be focusing on the relationship between science and our confessions of faith.

Evolution Sunday and Weekend is observed on the weekend nearest Charles Darwin's Birthday. We don't do this because Darwin was a saint, or because he had special spiritual knowledge that we need to pass on. But Darwin is important to our conversation, because he personifies the ongoing debate that has rocked our churches and society for decades if not centuries. Although the debate started long before Darwin – just ask Galileo – the publication of Darwin's *On the Origin of the Species* changed the conversation between science and theology forever.

I've been participating in this observance since its inception, even though I'm not a scientist, nor even well-trained in the sciences. I do believe very strongly, however, that this conversation has important implications for both church and society. Issues like climate change, stem cell research, end of life, and human sexuality all have both scientific and theological implications. If we're going to make good decisions about the future of our world, then we need to listen to both voices. This won't happen, however, if one side sees the other as the enemy. So, how can we have a fruitful conversation about science and faith, when the two sides seem so far apart? Well, that's why this observance was born!

As we consider how faith and science relate to each other, I'd like us to reflect on Job 38. But, before we get to our reflections, I'd like to read it again, this time from *The Message.*

> 1 And now, finally, God answered Job from the eye of a violent storm. He said:
> 2-11 "Why do you confuse the issue?
> Why do you talk
>   without knowing what you're talking about?
> Pull yourself together, Job!
> Up on your feet! Stand tall!
> I have some questions for you,
> and I want some straight answers.
> Where were you when I created the earth?
> Tell me, since you know so much!
> Who decided on its size? Certainly you'll know that!
> Who came up with the blueprints and measurements?
> How was its foundation poured,
> and who set the cornerstone,
> While the morning stars sang in chorus
> and all the angels shouted praise?
> And who took charge of the ocean
> when it gushed forth like a baby from the womb?
> That was me! I wrapped it in soft clouds,
> and tucked it in safely at night.
> Then I made a playpen for it,
> a strong playpen so it couldn't run loose,
> And said, 'Stay here, this is your place.
> Your wild tantrums are confined to this place.'

I should point out that the conversation that begins in these eleven verses goes on for two whole chapters. Aren't you glad we just read the opening round?

If you go back to the preceding chapters in this book, you'll discover that Job has a problem with the way the way God runs the universe. He doesn't curse God, but he does complain about the unfairness of his plight. He also doesn't appreciate the fact that his friends have accused him of being a sinner. Now, we get to watch God's response to Job's outburst. As we read this passage, it's almost as if God is trying to pummel Job into submission with unanswerable questions. I don't know about you, but I don't particularly care for this portrait of God. I find the God present in much of Job to be capricious, angry, self-centered, and a bit of a bully. And yet, I believe that we can find a word of wisdom in this passage that will speak to the questions of our day.

WERE YOU THERE? A MATTER OF PERSPECTIVE

As I read this passage, I thought of the song **"We're you there, when they crucified my Lord?"** It's not that the song speaks to the question at hand; it's just that opening line – were you there? – sounds like the question God asks of Job. You've questioned my fairness, so here are a couple of questions for you. You seem to have all the answers, so where were you in the beginning? You talk big, but do you have all the facts? When this barrage of questions ends two chapters later, Job had no choice but to answer, "No I wasn't there and so I can't answer all your questions."

Although it might seem as if God is beating up on Job, God's question does raise the issue of perspective. This passage reminds us that our vantage point as human beings is limited. We can't know everything there is to know about the universe. It's simply too vast. But that doesn't mean that there are no answers or that we should stop asking our questions.

I appreciate the point made by Daniel Harrell, a pastor who is trained in the sciences. He writes that science has an easier time dealing with the questions of nature, because, "science has more clear-cut boundaries than does theology." He goes on to say:

> Science limits itself to the natural, measurable world, while theology expands to include the immeasurable too. Everything science investigates is subject to scrutiny and testing,

but when it comes to God, our posture is to be one of deference and obedience.[1]

In other words, theology explores areas that ultimately require from us a confession of faith.

GET READY TO DEBATE

Our text speaks of perspective, but it's also a call to pursue the truth, no matter where it takes us. I realize that the ferocity of God's questions seems to shut down the conversation. If you look at the beginning of chapter 40, it appears that Job had decided that it might be better not to ask any more questions. And yet, hidden inside this series of seemingly unanswerable questions, is a challenge to pursue our questions no matter where they lead. God tells Job: "Gird your loins, be a man, stand tall, and answer my questions." Is this not an invitation to pursue the questions on our hearts and minds?

As I think about this question, I find myself turning to that statement of Augustine about "faith seeking understanding." As I understand Augustine, we shouldn't see faith as our answer of last resort. Instead, it's our starting point, upon which we build understanding of the things of God – and that includes nature. As Jesus reminds us, the Law of God calls us to love God with all our heart, soul, and mind. Faith, therefore, gives us the freedom to explore our doubts and questions without fear of what we might discover.

The New Atheists, such as Richard Dawkins and Daniel Dennett, talk about people of faith as if they don't think and they're gullible. Dawkins talks as if theology is simply a collection of old worn-out fairy tales. Unfortunately, there's some truth to the charge. The polls tell us that large numbers of Americans believe that the earth is only 6000-years-old and that evolution is a satanic plot. So, maybe the critics can be forgiven for the skepticism that there is another way of looking at things.

---

1   Daniel Harrell, *Nature's Witness: How Evolution Can Inspire Faith,* (Nashville: Abingdon Press, 2008), p. 67.

If we're going to have a fruitful conversation about the relationship of faith and science, then people of faith, like us, should respect science and its contributions to the conversation. We shouldn't act as if Genesis or Job offers a scientific explanation of the way things are. If we're willing to do this, then perhaps a skeptical scientific community will be willing to engage us in this important conversation.

A CALL TO HUMBLE ADORATION

In my mind, theology tells us a lot about the meaning and purpose of the universe. Genesis tells us that this world of ours is good, and that we've been entrusted with its care. Job reminds us that the universe is God's Temple, and we've been invited to worship the God who both creates and inhabits this Temple. What theology doesn't do is fill in the gaps of our science. If we use God to fill in the gaps, then if science finds an answer, then God's place in the universe becomes smaller. So, instead of looking for God in the Gaps, perhaps we can, as Stephen Barr suggests, find God present in the "beauty, order, lawfulness, and harmony found in the world that God had made." Or, as John Calvin puts it:

> "God [has] manifested himself in the formation of every part of the world, and daily presents himself to public view, in such manner, that they cannot open their eyes without being constrained to behold him." And, "[W]ithersoever you turn your eyes, there is not an atom of the world in which you cannot behold some brilliant sparks at least of his glory. . . . You cannot at one view take a survey of this most ample and beautiful machine [the universe] in all its vast extent, without being completely overwhelmed with its infinite splendor"[1]

While I acknowledge that there is also evidence of disorder and chaos in the universe, I also believe that nature does bear witness to

1   Stephen Barr, "The End of Intelligent Design," *First Things,* http://www.firstthings.com/onthesquare/2010/02/the-end-of-intelligent-design).

God's eternal presence and ongoing work of creation – whether it's a beautiful sunset or a glorious snow-capped mountain.

As we observe Evolution Sunday, may we again see science and faith, not as enemies, but as partners in an ongoing conversation about the world in which live and work and have our being. And as people of faith, who respect the findings of the scientist, may we also stop to give thanks to God for the wondrous gift of nature.

Preached: February 14, 2010 (Evolution Sunday)
Central Woodward Christian Church (Disciples of Christ)
Troy, Michigan

7

GIVE GOD THE GLORY

Psalm 96

Music has the power to stir our souls and enliven our hearts and minds. Whenever Handel's Hallelujah Chorus is played or sung, nearly everyone stands. They may even join in singing the chorus. It happened just the other day, when Pat concluded his recital with this very piece.

Why do we do this? Is it just habit or expectation? Or is it because this piece of music is so inspiring that we cannot take it in sitting down? What is important to point out is that the Hallelujah Chorus, like Psalm 96, calls forth from us, a declaration that God is sovereign, not just over our personal lives, but as the Psalmist declares, over "all the earth." And so we sing:

> "Hallelujah For the Lord God omnipotent reigneth, hallelujah"

And then, we proclaim:

> The kingdom of this world is become the kingdom of our Lord,
> And of His Christ, and of His Christ;
> And He shall reign for ever and ever . . .
> "Hallelujah For the Lord God omnipotent reigneth, hallelujah"

In this song of praise, we hear echoes of the biblical declarations of God's reign, declarations like the one found in another ancient hymn, one that Paul included in his letter to the Philippians. This hymn declares that the one who emptied himself of glory has been raised up by God,

So that at the name of Jesus every knee should bend, in heaven and on earth and under the earth, and every tongue confess that Jesus Christ is Lord, to the glory of God the Father (Philippians 2:9-11).

I don't know what instrumentation Paul imagined for his hymn, but I expect it carried a sense similar to that of the Hallelujah Chorus and the 96th Psalm. Ultimately, it doesn't matter if our songs are accompanied by mighty organs, simple guitars, or even no accompaniment at all. What matters is what comes forth from the heart as a declaration of allegiance, thanksgiving, and praise.

The 96th Psalm calls for us to sing to God a new song. The Psalmist invites us to join with the whole of creation in singing the praises of God, who is our creator. It is by classification an enthronement psalm, which acknowledges the reign of God, and in this case also declares the good news that God is at work bringing salvation, healing, wholeness, and hope to a world that is fragmented and broken. It evokes from us visions of God's splendor, which is reflected in the beauty of God's creation. And then, it closes by offering us promises of stability and justice.

AFFIRMING GOD'S GLORY AND GREATNESS

And so, at the invitation of the Psalmist, we come before the throne of God, singing a new song that declares before all the creation God's glory and greatness. In doing this we affirm that God transcends our boundaries and our lives. God is present with us and among us and even within us through the Spirit, but we are not God. Karl Barth speaks of God as being "wholly other." That may or may not be sufficient definition of God's being, but it is a reminder that when approach God, we stand upon holy ground.

When Moses went to the mountain to receive instructions for God's people, God reminded him that he stood on sacred ground and that he should take off his shoes. Here in this Psalm, we're directed to:

Ascribe to the Lord, O families of the peoples,
Ascribe to the Lord glory and strength.
Ascribe to the Lord the glory due his name;
bring an offering, and come into his courts (vs. 7-8).

Come into God's presence, bringing with you both words of praise that affirm God's greatness, and bring signs of your devotion, offerings that affirm your allegiance to the one who sits on the throne of heaven.

Enjoying God's Beauty and Splendor

Even as the Psalmist invites us to kneel before the Lord our Maker, the writer declares that "honor and majesty are before him; strength and beauty are in his sanctuary," and then invites us to worship God "in holy splendor." This is an invitation to enjoy the beauty and splendor that is reflected in God's creation.

Consider the wondrous beauty of God's creation, whether it's the dunes along Lake Michigan, the deep blue waters of Crater Lake, or the majesty that is Mount Shasta. Each of us can name a place that is so beautiful that we can't do anything except stand or kneel in awe. There are other expressions of God's splendor that come from within us, as we are invited to co-create with God things of beauty and grace. This invitation is written into our very being, for as Genesis reminds us, we have been created in the image of God. And so, it is our calling to bring forth beauty and splendor in the world. It might be music, such as we see displayed by the choir or the organ. It might be a piece of art or a poem.

N. T. Wright speaks of humankind being the reflection of God's "wise, creative, loving presence and power." God is enlisting us, in our very creation, "to act as his stewards in the project of creation." Therefore, Wright states that:

Every act of love, gratitude, and kindness; every work of art or music inspired by the love of God and delight in the beauty of his creation; every minute spent teaching a severely handicapped child to read or to walk; every act of care and nurture, of comfort and support, for one's fellow human beings and for that matter one's fellow nonhuman creatures; and of course every prayer, all Spirit-led teaching, every deed that spreads the gospel, builds up the church, embraces and embodies holiness rather than corruption, and makes the name of Jesus honored in this world – all of this will find its way, through the resurrecting power of God, into the new creation that God will one day make. That is the logic of the mission of God.[1]

We speak of ourselves as being a missional church. Therefore, as we create beauty we express God's mission by helping create a better world, a world in which God's name is honored and praised because there is joy and there is hope.

EXPERIENCING STABILITY AND JUDGMENT

As we reach the closing stanzas of this great Psalm, a Psalm that directs us to proclaim the good news of God's salvation, we hear words about judgment and stability. As we've been learning in the Wednesday studies, salvation isn't about being whisked away from this world by God. Instead, God's work of salvation is about making the world whole, and as we experience this wholeness – not perfectly of course – we have the opportunity to participate in God's work of healing that which is broken. It is, to quote Paul, our participation as ambassadors of reconciliation, even as God, in Christ, is reconciling us to God's self, so that we might experience the new creation (2 Cor. 5:16-21). Salvation has a partner, and that partner is judgment. Now, in our study, we've also been learning that God's judgment and justice aren't about punishment and condemnation. Although, God separates that which is good and honorable from that which is evil and dishonorable, God is not doing this in order to punish or condemn. God's judgment is designed to make things right so that there might be peace and

---

1   N.T. Wright, *Surprised by Hope,* (San Francisco: Harper One, 2008), pp. 207-208

good will on earth as in heaven. The Psalmist declares that God will come to judge the earth in righteousness and truth. If we trust that God is not just fair, but gracious and merciful and loving, then we need not fear God's justice. Instead, we can find in this message a word of hope, for God is not abandoning us or this world, but God instead is seeking to make things new.

Even as God promises to come and judge with righteousness and truth, we also hear a promise of stability. The Psalmist declares that the "World is firmly established and shall never be moved." Now, that doesn't mean that the earth won't experience quakes or other cataclysms. I suppose it's even possible that California could break off and fall into the sea, making Las Vegas a beach town. Rather than hearing this in a geological sense, perhaps we should hear it in the context of living in a mobile culture. It is an invitation to put our roots down into God's presence and entrusting our lives to the care of God. Jonathan Wilson-Hartgrove writes of "the wisdom of stability," and speaks of stability as being "a commitment to trust God not in an ideal world, but in the battered and bruised world we know. If real life with God can happen anywhere at all, then it can happen here among the people whose troubles are already evident to us."[1]

With this promise of stability as our anchor in this world, may we join together with the seas and the fields and the forests, and sing for joy before the Lord, declaring that God is glorious and great. Yes, let us sing: "To God be the glory, great things he hath done!"

Preached: May 30, 2010 (Trinity Sunday)
Central Woodward Christian Church (Disciples of Christ)
Troy, Michigan

---

1   Jonathan Wilson-Hartgrove, *The Wisdom of Stability*, (Brewster, MA: Paraclete Press, 2010), p. 24.

8

In the Beginning . . .

Genesis 1:1-5

"The end of something is better than its beginning" (Ecclesiastes 7:8 CEB).

I thought you'd want to hear this word from Ecclesiastes, since we're moving into a new year. Beginnings are important, but endings are even more important. A few years ago the Lions won all their preseason games and everyone expected good things, and then they lost the next sixteen in a row. This year, the Lions had an up and down season, but they ended up in the playoffs – that was a much better conclusion.

Each of us has a story of beginnings to tell; what we don't know is how things will turn out. My own life began on March 3, 1958 in Los Angeles. Five years later, I began my formal schooling as kindergartner in Mt. Shasta. From then on, for the next seventeen Septembers, I would begin a new school year. After taking off two Septembers, I restarted school in January 1982, when I began my seminary career. Of course I didn't just start school; I also began a new phase of life on a summer day in July when Cheryl and I were married. There was another day of beginnings in June of 1985, when I was ordained. Then, there was that day in April of 1990, when I became a parent. These are just some of the beginnings of my life experiences. I've had some endings, but there are still many of these beginnings that have yet to reach an ending.

Genesis 1 begins with the words, "In the beginning," a phrase that is picked up by the Gospel of John, which declares:

> "In the beginning was the Word, and the Word was with God, and the Word was God. The Word was with God in the

beginning. Everything came into being through the Word, and without the Word nothing came into being (Jn 1:1-3a CEB).

The Gospel of Mark starts with the words: "The beginning of the good news of Jesus Christ, the Son of God." And in Revelation Jesus declares: "I am the Alpha and the Omega, the Beginning and the End" (Rev. 21:6), while Paul writes:

> So then, if anyone is in Christ, that person is part of the new creation. The old things have gone away, and look, new things have arrived! (2 Corinthians 5:17 CEB)

A time of beginnings, whether new or not, is a time when we begin with a clean slate. Whatever happened in the past, for good of for bad, is in the past. Now, however, is the time to move into the future, embracing all the opportunities the future presents. And if God promises to be with us in the beginning, then surely we can expect God to be with us until the end of all things. Although the end is better than the beginning, the writer of Ecclesiastes cautions us to be patient.

Before we get too focused on the end of things, perhaps we could return to the beginning of the story – with the opening lines of Genesis 1.

Do you hear a difference between this more traditional translation of the opening line of Genesis – "In the beginning God created the heavens and the earth" – and this more recent translation: "When God began to create the heavens and the earth?"

Did the universe emerge fully formed in a single moment? Or is creation a process that is even now unfolding? Is God finished with the work of creation and living in retirement? Or is God still at work bringing order out of chaos and bringing light into the darkness?

Since I'm not a Deist, I'll cast my lot with the God who is still at work bringing order out of chaos and light into the darkness. Because we live in a world where both chaos and darkness are still present, I find hope in the promise that God is still at work. What is, is not the final word.

We have experienced the Alpha, but not the Omega.

If we step back to look at the full picture of creation – from beginning to end – perhaps we can see why God declares the creation to be good. It is full of potential. But, as we look closer, we see that things are not yet complete. Disorder is still present and darkness remains. Where is God in the midst of this disorder and chaos? Is God too weak to bring order to creation? Or is the power of God expressed differently than we've often been led to believe?

May we say that God is at work, but not as the omnipotent miracle worker, who reaches and fixes things when they're broke, but the one who comes to us in what is revealed in and through the cross of Jesus?

And where do we fit in the story? If you continue reading Genesis 1, you will come to the creation of humanity, and you will hear God give humanity stewardship over the creation. In doing this, God entrusts us with co-responsibility for this creation. Mixed into this story, however, is the continued presence of chaos and darkness. The biblical story gives us two words – God's creation is good, but evil is real and it is present in our midst. As we read on we discover that humans can get caught up in this darkness, but we also hear the promise that God has made provision in Christ to bring us back into the light. This is the good news – the word of liberation.

We know that in this time and place there are those who are experiencing a moment of darkness. May we become light bearers in their moment of darkness. There are also those who are experiencing chaos. It could be the crisis of foreclosure, the loss of a job, or simply a loss of hope. May we be instruments of God's ordering of life.

As we ponder this calling to join with God in bringing order and light into the world, may we also remember that even as we seek to respond to this invitation, we too need light and order in our own lives. We are not free from complicity in the chaos and darkness of this life, for the ending is not yet upon us.

Perhaps we can hear a word of guidance in Mark's story of Jesus' baptism by John. In this account, Jesus joins with the throngs of people coming to John to be baptized, as a sign that the desire to change their hearts and lives. Jesus submits to this baptism, identifying himself with the sinners of this world. But, as Jesus emerges from the waters of the Jordan, the heavens open, the Spirit of God descends like a dove upon Jesus, and then a voice from heaven declares: "You are my Son; whom I dearly love; in you I find happiness."

Yes, there is happiness to be found in the one through whom and in whom order and light comes into the world. That same Spirit that descends upon Jesus is the same Spirit who hovered above the waters at the beginning of creation, when life itself began. This same Spirit fell on the Day of Pentecost upon the disciples, empowering them to share with God in this work of bringing order and light into the world.

Although the end may be better than the beginning, may this be a day of New Beginnings.

Preached: January 8, 2012 (First Sunday after Epiphany)
Central Woodward Christian Church
Troy, MI

9

How Do We Hear the Voice of God? (Reason)

Note: This sermon was preached as part of a series that included messages on Scripture, Tradition, and Experience, as well as Reason.

Acts 17:16-21

I think we've established over the past few weeks that even if God doesn't normally speak to us in an audible voice, we can still hear the voice of God. We just need help. There's Scripture, of course, which we often call the Word of God, and it is normally our starting point. After all, we read from Scripture every Sunday as part of worship. But as the Gospel of John reminds us, Jesus, not the Bible, is the Word of God in the flesh. Although Scripture seems to be a central way in which God speaks to us, is it the only way we can hear God speak?

We started to answer this question last Sunday with a conversation about Tradition, which is the ongoing story of God's involvement in our world, beginning with Creation and continuing to this day. Tradition is an important voice, but perhaps there are still others that might speak to us. If so, could Reason be one of those ways in which God speaks?

In planning worship this week I discovered that there aren't many songs and hymns that celebrate Reason. I also couldn't come up with any great Broadway songs as a follow-up to last week's wonderful song from the Fiddler on the Roof' – Tradition. I did think about suggesting to Pat that he might want to play the original Star Trek Theme as the Prelude, but then Logic suggested that might not be a very reasonable idea.

So, even if we don't have a great Broadway tune to celebrate it, can God speak to us through Reason? Could resources like philosophy and science be ways in which God's voice might be revealed to us?

Although the Founders of the Disciples Tradition believed Scripture was the primary way through which God speaks to us, they also believed that the Christian faith should be a reasonable one. They were deeply attracted to thinkers like John Locke, who believed that Truth is largely self-evident. We just have to open our eyes to this self-evident Truth. You'll find a similar attitude present in the American Declaration of Independence, where Thomas Jefferson, who was a Deist, wrote:

> We hold these truths to be self-evident, that all men are created equal, that they are endowed by their Creator with certain unalienable rights, that among these are life, liberty and the pursuit of happiness.

How do you know that you've been endowed with these inalienable rights? Reason tells you that this is the Truth. It's self-evident.

Although Alexander Campbell believed that some things having to do with our relationship with God lie beyond the bounds of unaided human understanding, he also believed that religious Truth shouldn't conflict with Reason. Like Joe Friday, he believed in the facts, and you will find the Facts revealed in Scripture, and this revelation should not conflict with Reason. He wrote:

> Indeed, faith, Divine faith, is the conviction or evidence of things not submitted to our senses. But in no case does it conflict with the true and proper constitution of the human mind—nor with the power, wisdom, and goodness of God as developed in creation.

So, if you hear someone say that Scripture and science are in conflict, and that they'll go with Scripture over science, Campbell might respond by telling this person to better check their facts, because they must be missing something.

It shouldn't come as a surprise to anyone who knows me that I value things intellectual, which may be why I've been attracted to Campbell's vision of a reasonable faith. And I'm not alone in this. I've had many conversations with members of this church about just this "fact." I expect that many of you would join me in agreeing with Galileo, who said: "I do not feel obliged to believe that the same God who has endowed us with sense, reason, and intellect has intended us to forgo their use." And since Scripture commands us to Love God with our entire being, which includes heart, soul, strength, and Mind, I believe that it's probably okay to pay attention to the Mind.

Just to make sure I was on the right track, I decided to do a bit of online research, and I did a search to see what the Bible had to say about this topic. I have to admit that some of the passages I uncovered proved to be a bit discouraging. Consider what Ecclesiastes 1:18 has to say: "In much wisdom is much aggravation; the more knowledge, the more pain" (CEB). Maybe the writer of these words of wisdom had just finished an exam, but Paul said something similar. Although Paul told the Romans that the things of God should be plain to us, because God is revealed to us through the things God has made (Romans 1:19-20), in 1 Corinthians Paul insists that the foolishness of God is greater than human wisdom (1 Corinthians 1:18-25). So much for Galileo's vision! Maybe Reason is a dead end!

Then we come to the story we read together just a moment ago. Paul had gone to Athens, which in Paul's day was a university town, something like Ann Arbor, though without the football stadium. It seems that the Athenians loved to spend their time doing nothing but talk philosophy and theology. These are Reasonable people! They love to talk about new ideas. But it's not just the residents of Athens who love to talk about such things, even the visitors – people like Paul, for instance – loved to join in the conversation.

Luke mentions Paul's debates with the Stoics and the Epicureans, but there were a lot of other schools to engage as well.

Besides the schools of philosophy there were synagogues, temples, and shrines. In fact there were so many that Paul seems to have been deeply distressed by what he found there. But, he also found a shrine dedicated to the "unknown God." Since no one had claimed this shrine, Paul decided he would fill in the blank, and preach the God of Jesus as this "Unknown God."

So, after spending some time preaching in the synagogues Paul went out and stood on his soap box and started giving a lecture. Remember this is a university town so people might have enjoyed listening to lectures and engaging in debates. The people even dragged Paul before the Council so that he could be interrogated. You see, many of the people found Paul's message to be rather unreasonable. It wasn't all that self-evident, especially Paul's message about the resurrection.

You see, many Greeks believed that the body is a prison, and the goal is to free yourself from that prison. So why would you want to experience resurrection if that meant continuing your bodily existence? Although the mind and the soul are good, the body is simply a hindrance to our ability to enjoy the spiritual.

Although I don't enjoy admitting this, it appears that Paul's stay in Athens was not a success. He didn't plant a church nor did he gain many converts. It seems that the Athenians just didn't think the Gospel was very reasonable. If that's true, then why should we believe that God would speak to us through human reason? Why should we take Galileo's word over that of the writer of Ecclesiastes?

And, even if Campbell, Locke, and Aquinas value wisdom, could Martin Luther have been right when he said:

> Reason is a whore, the greatest enemy that faith has; it never comes to the aid of spiritual things, but more frequently than not struggles against the divine Word, treating with contempt all that emanates from God.

After listening to Luther, maybe I should just sit down and forget about finding any help from Reason, except I like what Galileo had to say about God creating us with Minds, expecting us to use them.

I think that Luther was a bit like Dr. McCoy on Star Trek. He's well educated, but he's a passionate sort of person. Dr. McCoy seems to rely a lot on Emotion, while his friend, Mr. Spock values Logic and seeks to suppress Emotion. As you may know there's a third member of what we might call the Star Trek trinity. That person is Captain Kirk who always finds himself listening to these very distinct voices – Spock's Logic and McCoy's Emotion. Like Kirk we also listen to both voices. So, as we seek to know the way of God it would seem that we must balance these two voices. We are, you might say, people with both hearts and minds. When we overemphasize one at the expense of the other, is it possible to truly hear the voice of God?

So, do you think ours should be a reasonable faith? Do you think that God might reveal some of what God would have us know and understand through science and philosophy? Even if some of the things of God may appear at times to be foolishness to some, does accepting the Gospel mean losing our minds?

As we ponder this question, perhaps a word from James will offer us guidance: He writes that "anyone who needs wisdom should ask God," and "wisdom will certainly be given to those who ask." (James 1:5).

I think Galileo would agree with James – as would John Locke and Alexander Campbell, who both believed that faith and reason can get along!

Oh and just so we don't leave out Dr. McCoy, next week we'll talk about how God might speak to us through experience.

Preached: May 18, 2012
Central Woodward Christian Church (Disciples of Christ)
Troy, Michigan

Essays

1

Finding a Middle Ground
in the Evolution Wars

Perhaps like me, you find the battles being fought in school board meetings and science class rooms across the country, distressing and even disturbing. The actions of the Dover, Pennsylvania school district in mandating mention of "intelligent design" in biology classes, and the Kansas state school board decision to downgrade evolution in its state science standards have been much in the news. It would seem that the scientific establishment is under siege by the religious community, or is it the reverse? Is the science class room proving to be a stumbling block to the faith of our young people? Dover and Kansas may seem far away from Lompoc, but the debate goes on even here, in churches and classrooms, living rooms and in barbershops. God or Darwin, these seem to be our only choices, but are they?

This debate is not new, of course. Just think of the Scopes trial of the 1920s, the Supreme Court case in the 1970s that overturned an Arkansas ban on evolution, and the suppression of Galileo by the church. Intelligent design in and of itself is not an enemy of science, but as its proponents begin to argue that it is an alternative to evolution then confrontation is sure to erupt, and it has.

There is of course, a middle ground between two extremes – an atheistic scientific naturalism and an apparently anti-science religion. It took some time, but today most Christians aren't threatened by Galileo's discovery that the universe doesn't revolve around the earth. Darwin stirred up controversy a century and a half ago with his theory of natural selection, but from that day to the present there have been many theologians who have tried, I think successfully, to reconcile Darwin and faith. It is unfortunate that in the current debate, this middle voice seems to be drowned out by the partisan nature of the debate.

I write as a pastor and theologian and not as a scientist, but I am not afraid and my faith is not threatened by science. Science may force me to rethink some of my conclusions and beliefs, but I am comfortable with the theory of evolution. I find it hard to refute a theory that receives almost unanimous support from the scientific community. This theory is the foundation of many medical discoveries that benefit our everyday lives – why else would scientists be able to test medicines on mice if there weren't some connections between us.

The proposed alternative – "intelligent design" – has received political support from President Bush and many evangelical Christians, but this is not a scientific theory as some allege. It is in fact a philosophical statement, which insists that gaps in evolutionary theory and the complexity of the universe require an "intelligent designer," who may or may not be God. What proponents of intelligent design have yet to offer is a compelling and testable scientific theory that explains the facts on the ground.

The question of design is not a bad one. Of course, recent earthquakes and hurricanes can raise questions about design, even as complexity raises questions of an unguided evolution. Perhaps, we would be better served if science and religion moved from debate to dialog. These two communities need not be adversaries.

These communities come into conflict when they speak of things of which they are not equipped to speak. Science ultimately cannot speak of God, and we in the religious community get in trouble when we ask of our faith traditions the wrong questions. Genesis does not offer scientific theory, but it does offer a wonderful statement about the meaning and purpose of creation, something scientific theory is less equipped to handle. At the end of each of six creative acts, God says "it is good." This world, in which we live, is good and it should be cared for and nurtured and enjoyed. Creation, Genesis 1 and 2 suggest, is the context of a relationship between a loving God and human beings. On this, science is relatively quiet.

It is healthy to debate the merits of science and faith. The questions of intelligence and design and their relationship to science are worth pursuing, but perhaps they can be taken up in a venue other than a science class room or a school board meeting. The consequences of continuing the battle will be for both science and faith. Is it any surprise that in this fractious climate that America is falling behind the rest of the developed world in science? At the same time, is it any wonder that many people sear off faith in God, because it would seem that to embrace God is to deny science? And yet, there is a middle path that can lead to fruitful dialog, if only we will choose to take it.

Lompoc Record,
Faith in the Public Square
October 16, 2005

2

Are You Observing Evolution Sunday?

Charles Darwin's return from his voyage on the H.M.S. Beagle turned the world upside down. Not only did his discoveries revolutionize biology, they nearly put God out of business. Or, so it seems.

Darwin's proposal that natural selection should replace design as the mechanism of creation continues to provoke uproar a century and a half after the publication of Origin of the Species. Partisans, pro and con, argue over whether human beings descend from a very nonhuman common ancestor and whether evolution is by definition blind and purposeless.

Scientific consensus does insist that evolution, by some means akin to natural selection, explains the world that exists. Despite the evidence offered by the scientific establishment, half if not more of Americans not only reject evolution, they insist that life exists today pretty much as it always has. In the popular mind, it would seem that God has triumphed over science. Whereas two decades ago, the watchword was scientific creationism, today we argue about Intelligent Design. Intelligent Design may not be creationism under a new name, but it does seek to undermine evolution. It does so by insisting that the world has irreducible complexities that require a designer and cannot be explained by random developments.

Darwin did end up an agnostic, but from his day to the present many religiously-inclined scientists, along with devout theologians, clergy, and biblical scholars have sought to embrace both evolution (including natural selection as its mechanism) and the notion that God is the creator. They seek to hold together the seeming randomness of the natural world with a sense of design. Whether it is called "theistic evolution" or "evolving creation," the point is, God and modern science need not be mutually exclusive entities.

Today, more than 400 Christian congregations from across the nation are observing "Evolution Sunday" as a way of reclaiming a healthy relationship between science and Christian faith. It follows upon "An Open Letter Concerning Religion and Science" that has garnered more than 10,000 signatories, mine included. Signatories include local clergy, biblical scholars, theologians, philosophers, bishops and denominational heads, as well as university and seminary presidents. They represent religious communities as diverse as Roman Catholic, Mainline and Evangelical Protestant churches, along with a few Unitarians. Some names are well known, others are not.

Each signatory affirms the premise that evolution and Christian faith can live in harmony, as long as each respects the competency of the other in its own field of inquiry. It allows for conversation and cooperation, but it rejects the either/or choices so often presented to the community. The letter affirms that "the theory of evolution is a foundational scientific truth, one that has stood up to rigorous scrutiny and upon which much of human knowledge and achievement rests" It also affirms the biblical witness that God is the creator. In doing this, the signatories agree to build a bridge between two seemingly opposite ventures. To do this we must face the question of biblical interpretation. If we must read Genesis as a modern historical and scientific statement, then the bridge will fall. But one need not read Genesis in a literal fashion to be faithful to its message, which is, all that what exists is from God and it is in its own way very good.

"Evolution Sunday" may seem like an odd addition to the liturgical calendar, but it is not as strange as it seems. By observing this event, the church brings science back into conversation with theology. It recognizes that each discipline looks at the world from a different point of view. As a Christian I affirm God's intimate involvement in the universe. I confess God as Creator and stand in awe of God's handiwork, even while recognizing that such beauties as Crater Lake and the Gaviota Coast result from natural causes.

This Evolution Sunday I join the Psalmist in declaring that the "heavens are telling of the glory of God; and the firmament proclaims his handiwork" (Ps. 19:1), while affirming the theory of evolution as a scientifically valid explanation for how things have come to be. Today is the day to restart the conversation between science and faith.

Lompoc Record
Faith in the Public Square
February 12, 2006

3

Building Bridges –
Observing Darwin's Birthday in Church

On February 12th my congregation in California was one of four hundred plus churches across the nation that observed Evolution Sunday. Days before this event, which coincided with Charles Darwin's 197th birthday, the Discovery Institute, the Seattle-based proponent of Intelligent Design, opined that our observance was "the height of hypocrisy." Discovery Institute officials charged us with sharing in a bit of "old time Darwinist religion." Yes, I gave a "pro-evolution" sermon, but I'm not sure what this "good old Darwinist religion" is, and I would object to this characterization of my efforts. I expect the same would be true of other preachers who chose to observe this event. Yes, we affirmed evolution as a scientific fact, but I wouldn't call this Darwinist religion nor would I call it hypocritical.

While I can't speak for my fellow preachers, nor the 10,000 plus signatories to the Clergy Letter Project's "Open Letter on Religion and Science," my sermon, and probably many others preached that day, affirmed evolution and recognized Darwin's contribution to the theory, but it also celebrated God the creator. I sought to build a bridge between Christian faith and the science that explains the universe we live in. Whatever the challenges and modifications made over the last century and a half to the theory of evolution and Darwin's theory of natural selection, evolution remains the acknowledged explanation of earth's development and the diversity of its *flora* and *fauna*. While some proponents of evolution insist that it is a blind and purposeless process, this is not true of all its proponents, especially those of us who are theists.

I had my congregation observe Evolution Sunday because I'm concerned about the growing divide between the religious and scientific communities. I also spoke on this topic out of concern for the gospel of Jesus Christ, which can appear irrational and anti-intellectual in light of the vitriol expressed by some Christians against evolution.

Despite the overwhelming evidence presented in our nation's public schools and in myriads of nature programs on TV, nearly half of Americans reject evolution. School boards and legislatures across the country are changing curriculum standards, often under pressure from religiously-motivated groups (despite disclaimers from the Discovery Institute). Whether or not Intelligent Design is creationism under a new name, its proponents seek to undermine evolution by "teaching [a] controversy" where no controversy exists.

As for the Darwinist campaign to use religion to promote evolution in the schools, I'm neither aware of it nor part of it. Evolution Sunday simply gave me the opportunity to speak of the harmony that can exist between theology and science if each will respect the competency of the other in its field of inquiry. Conversation and cooperation are essential, but they can't happen as long as partisans on both sides of the issue propose either/or positions. The "Open Letter on Religion and Science," which I signed, affirms that "the theory of evolution is a foundational scientific truth, one that has stood up to rigorous scrutiny and upon which much of human knowledge and achievement rests," and that this statement is compatible with the biblical witness that God is the Creator.

Building a bridge between biblical faith and evolutionary science, which the letter calls for, requires us to face the question of biblical interpretation. If Genesis must be read as a modern historical and scientific statement, then the bridge will fall. But, we needn't read Genesis in such a literal fashion to be faithful to its message, which is that what exists is from God and that this creation is very good.

As a Christian, I affirm God's intimate involvement in the creation of the universe, but I reject a "God of the Gaps" solution.

I also recognize that such beautiful examples of God's handiwork as the Grand Canyon, Crater Lake, and humankind can have natural explanations. Therefore, on Evolution Sunday, we joined the Psalmist in declaring that the "heavens are telling of the glory of God; and the firmament proclaims his handiwork" (Ps. 19:1).

Ponderings on a Faith Journey
February 24. 2006

4

LETTER TO A CHRISTIAN NATION –
A RESPONSE

Ninety percent of Americans, maybe more, believe in God – or at least we believe in some kind of divine being/entity. The number of avowed atheists is thought to be about 6 percent of the 300,000,000 Americans. That's not a large number. But the truth is the number of those who embrace some institutionalized form of religion is probably fairly substantial.

So I come to Sam Harris's best seller, *Letter to a Christian Nation*, (Knopf, 2006). I just finished reading this little book by the author of another best seller *The End of Faith* – a book I haven't read. This is a hard-hitting no holds barred, no prisoners taken, broadside against religion, and Christianity and Islam in particular. In the mind of Sam Harris the best thing that could happen to the world is to see religion eradicated. He holds out no hope for moderate or liberal versions of religious faith – they simply provide a cover to the extremists who are a danger to the world.

This book of course builds a straw man and effectively demolishes it. To Harris, there is only one true Christian, and both he and they agree on what that means. Christianity is by definition, irrational, obscurantist, anti-intellectual, and given to violence (I agree some forms are given to such things). Whatever good can come of religious life is more than outweighed by the bad things it produces. True Christians believe that the Bible is the Word of God and take it completely literally (even though a goodly number, myself included, seek to read the Bible critically and recognize that not everything needs to be taken literally). True Christians believe that Jesus is coming back soon (*Left Behind/Late Great Planet Earth*) and that anyone who doesn't believe as they do, is going to hell (though many Christians take a much more nuanced perspective – even embracing forms of universalism). That I don't recognize myself in his description shouldn't surprise me, so he says, because well, I'm really not a Christian.

There are some real absurdities here, but in spite of the often insipid and strange stereotypes (such as a discussion of biblical prophecy that wonders why the Bible if it's truly the word of God doesn't provide detailed instructions/information on really important scientific data), the book may prove useful. It is a reminder that religion, and Christianity in particular, has its dark side. He reminds us, usefully, that persons of other faiths, such as my Muslim friends, look at me in the same way I, as a Christian, have looked at them. We are equally committed to our position and believe the other is destined for hell because of what they believe. Simply to declare that the Bible is true and therefore if its words are not believed one is going to hell doesn't really prove anything.

He raises important questions about religion and science. It's saddening to me as well that so many Americans have abandoned the scientific consensus about evolution, preferring to believe varieties of creationism or design theory. This rejection of science extends to other issues, such as the APA's findings on homosexuality, and climate change.

Harris, who apparently holds a degree in philosophy from Stanford, believes that religion might have had an evolutionary benefit – a glue to bind developing society together – that glue is no longer needed in a rational and civilized world.

Do I accept his conclusions? No, I don't accept his stereotyped version of Christianity as true to the mark, but that's not the point. There are plenty of Christians, and religious persons of any number of traditions, who fit the stereotype. But, the point is simply this – though 90% of Americans believe in God, tons of people are intrigued enough to pick up Sam Harris and Richard Dawkins (*The God Delusion*) and imbibe what they have to say. This tells me that there are a whole lot of people who say they believe in God, but aren't quite sure. I expect that Sam will get a few converts, maybe a whole lot of them.

I'm not convinced by his diatribe, but I am challenged by it. I'm not convinced that I'm doing something irrational – if I am then a whole lot of very intelligent and thoughtful individuals are as irrational as I am. Several centuries back, Friedrich Schleiermacher wrote his *Speeches to the Cultured Despisers* – Harris I think is one of those cultured despisers. Schleiermacher turned to religious feelings – feelings of absolute dependence on God as the foundation of his defense. Whether that defense will work this time, I'm not sure. But, I also know that Harris isn't the first to raise these questions and won't be the last. But instead of assailing him for raising the questions, perhaps we would be well served by considering them for ourselves.

Ponderings on a Faith Journey
October 21, 2006

5

DARWIN MATTERS

Today, at the height of scientific knowledge, we ironically face a skepticism of science that in many ways is rooted in religious challenges. And nowhere is this skepticism more rabidly evident than in the debate over evolution.

First it was "scientific creationism," and then more recently "Intelligent Design," that has challenged the science behind the theory of evolution. The problem is that the challenge, though often disguised (especially within the Intelligent Design camp), is religiously based. Proponents of creationism and ID have played with and manipulated public perceptions of science that does harm both to science and to faith.

Because biblical literalists and proponents of "Intelligent Design" have captured the attention of the media and provided fodder for atheists such as Sam Harris and Richard Dawkins, it might appear that there are only two choices: atheism and science or belief and biblical literalism. The truth, however, is there are other ways of looking at these issues that are not so black and white, more nuanced and friendly both to science and to faith.

Emerging from this vast middle ground is something called Evolution Sunday. Now one might ask, why would a church observe something called Evolution Sunday, an event that coincides with Charles Darwin's birthday and would seem to be the proper province of a humanist society or a scientific organization? Such an assumption might seem reasonable, and yet hundreds of religious communities from across the country are participating in what has become an annual event.

For the second year in a row, congregations are stepping forward and affirming the view that the planet and its human inhabitants require our support. Pastors like me are recognizing that we have a voice that needs to be heard, if for no other reason than that important scientific discoveries could be delayed or dispensed with by religiously motivated opponents to science.

Of course, such an effort is not without controversy or opposition. Participants in last year's observance were accused by the Discovery Institute, the leading institutional proponent of "Intelligent Design," of involvement in "old-time Darwinist religion." Calling it the "height of hypocrisy," the anti-evolution organization accused organizers of using religious folk to defend a discredited "theory" of evolution. This was, to them, a Darwinist plot to confuse the issue by using religious voices to support the teaching of evolution in public schools while condemning anti-evolution efforts as being religiously motivated. Indeed, Jonathan Wells, a Discovery Institute member, using a Yale Daily News op-ed piece, recently accused Evolution Sunday participants of using a "bait and switch ploy," to slip "Darwinism" into schools by sugar-coating it as a benign scientific theory with no religious implications.

Participants in Evolution Sunday have been accused of taking part – perhaps unwittingly – in a grand ruse or conspiracy to introduce bad science and atheistic ideology into our schools. Wells insists that the vast majority of Americans reject Darwinism–and thus the reigning theory of evolution–because they see through the science and ideology.

I, however, am not one of them. I have not been duped by a Darwinist conspiracy. I have nothing to gain from such participation, except that it arises from a concern for our planet, and from a concern for the reasonableness of my faith profession. Not being a scientist, I cannot speak to the intricacies of Darwin's theory of evolution by natural selection. Yet a common sense look at the issues suggests that Darwin might be right. Consider for a moment the impact of evolutionary theory on modern medicine. Medical researchers depend on this principle as they test new medications on mice and other animals. If there were no relationship between these species, such experiments would be impossible. And yet, we benefit every day from discoveries that are rooted in evolutionary theory.

As a person of faith who seeks truth, wherever it leads, I have been convinced that the vast majority of scientists must be on to something. In making this affirmation I do not have a secularist agenda, for I do believe in God the Creator. But science must inform my understanding of God's creative ways. From Darwin's time to the present, good Christian theologians have been able to reconcile the two, and not just liberals, but even conservatives such as Benjamin B. Warfield. And why? Again, there is a commitment here to following facts where they lead.

Nearly 600 churches are planning to use their worship services to address this issue. Each in our own way, we're declaring that evolutionary science and faith are compatible. We seek to address the fears of those who believe that evolutionary theory is a threat to faith and that if allowed to stand alone in our schools, will erode belief in God. In response to these fears, we suggest a way to build a bridge of understanding. Although the Evolution Sunday project is relatively new, the need for religious people to address this issue isn't, and in each new generation theologians and religious leaders have stepped up to deal with the issue.

"Evolution Sunday" is the brainchild of Dr. Michael Zimmerman, Dean of the School of Arts and Sciences at Butler University, and formerly an administrator and science professor at the University of Wisconsin, Oshkosh. Several years ago, Zimmerman wrote an open letter regarding the compatibility of religion and science, a letter that has now garnered over 10,000 clergy signatures. Among those who have chosen to sign this letter are well-known figures in theology and biblical studies. The majority, however, are local pastors from a wide range of denominations, from evangelical to Unitarian.

I would suppose that most of us who have signed the letter were attracted to the project because it offered a middle way between two unattractive ideologies–Intelligent Design and a secularist scientism that leaves no room for divinity or spirituality of any sort. The suggestion that these clergy and churches are hypocritical and captives of some sort of "Darwinist religion" because they are trying to build a bridge between two seemingly disparate ideas has proven a bit bewildering to me. Why must recognition of evolution as scientific truth by religious leaders be equated with Darwinist religion? In fact, what is Darwinist religion? Those who make the charge never offer a compelling definition. They simply suggest that Darwinist religion is something akin to atheism or scientific materialism. Yet whatever the critics may think of "Evolution Sunday," the worship service that was held last year on February 12, 2006 at First Christian Church (Disciples of Christ) in Lompoc, California, celebrated neither atheism nor scientific materialism.

The sermon that day took Genesis 1 as its text and celebrated God's creative acts, while at the same time giving credit to the discoveries of science, which point toward an evolving creation. The hymns that day didn't sing the praises of Darwin, but of God the Creator. The congregation didn't even sing happy birthday to dear old Charles, nor were prayers offered in his name. The service, including the sermon, was geared to building a bridge between Christian faith and modern science. As a preacher, I staked a claim in the debate and refused to capitulate to those who want us to make either/or choices—God or Darwin. We will be doing the same this year as we observe Evolution Sunday on February 11th (the day before Charles Darwin's 198th birthday.)

Critics of evolution parade before an unsuspecting public the many challenges and modifications that have occurred over the years to unhinge theory of evolution, and Darwin's theory of natural selection in particular. But as yet, they have failed to present an alternative theory beyond an appeal to a god of the gaps—something they call "irreducible complexity." Such an appeal suggests that when we reach a seeming dead end, we should give up our search for an answer and just chalk it up to the designer (who might or might not be God). But that's not a good a way of doing science—giving up before you come up with a good answer. So, until otherwise proven, evolution by natural selection remains the acknowledged scientific explanation of this world of ours. What other adequate natural explanation is there for the great diversity of animal and plant life in our world?

I chose to have my congregation observe Evolution Sunday because I am concerned about the widening divide between the faith and scientific communities. If we reject science, then we will have no say in resolving the ethical dilemmas that science often presents to us. We may also end up hindering scientific discoveries that would benefit humanity. I also speak of this out of concern for the gospel of Jesus Christ. The heightened rhetoric against evolution has surely made the gospel seem to many observers irrational and anti-intellectual. It places unnecessary barriers in the way of those who might look for a place of healing, grace, and transformation, people who are not willing to abandon their right to think for themselves. It also keeps believers from participating in the scientific revolution of our day. If, as most surveys suggest, upwards of 90% of Americans believe in God, then the possibility that faith and science are incompatible will hinder America as it moves into the future.

From my perspective, it is unfortunate that Christians have contributed so much to the growing rejection of evolution in America, and I'm afraid that this trend may partially explain why so many young people today show no interest in the sciences. Despite the evidence presented in public school classrooms across the country and through myriads of nature programs on TV, nearly half if not more of the American populace rejects the evidence offered by the scientific community–something Jonathan Wells gleefully celebrates. School boards and legislatures from across the country are changing curriculum standards and either rejecting or de-emphasizing evolution. Proponents of Intelligent Design seek to undermine evolution by "teaching [a] controversy" where no controversy needs to exist – except perhaps in a popular mind that is being formed by this rhetoric.

Critics of religious moderates and liberals claim that we give cover to Fundamentalists. In observing Evolution Sunday we offer our response to their challenge. And to those who want to believe and yet find it impossible to reject the findings of modern science, we offer a way forward. Of course such an effort requires that we face the issue of biblical interpretation and face the challenge of our theological inheritance. It might require of us a modification of our theologies, but it does not require of us a rejection of faith in God the Creator.

As a Christian, I affirm God's intimate involvement in the creation of the universe, but I also recognize that such beautiful examples of God's handiwork as Crater Lake and Mount Shasta (I grew up in close proximity to both), the Grand Canyon, and the human bodies have their natural explanations. Both the scientist and the theologian describe the same phenomena, but they use different vocabulary and tools to do so. If we are willing to recognize and affirm the contributions of both the scientist and the theologian, there can be a meaningful and profitable conversation—a conversation that has important consequences for human society and the planet we inhabit.

On Evolution Sunday, February 11, 2007, we at First Christian Church (Disciples of Christ) of Lompoc will acknowledge and affirm the work of the scientist and the wisdom of an evolving creation by joining with the Psalmist in declaring that "The heavens are telling of the glory of God; and the firmament proclaims his handiwork." (Ps. 19:1).

Published at SoMA Review
February 10, 2007

6

When Did I learn about Evolution?

I was talking about evolution with my father-in-law. Like many Americans he'd rather choose Genesis over what the scientists teach. He said to me: well, we didn't learn about that stuff when I was in school. That got me to thinking: Did we talk about evolution when I was in school? Now, I took biology in ninth grade, way back in 1972-1973. My biology teacher was beloved by the students. He got leeches and other wonderful delicacies out of the local canal for us to look at and I think dissect. But my teacher was also an Evangelical and the sponsor of our student-organized Bible club. I don't remember talking about evolution, one way or another back then. So, maybe I didn't learn anything about it. And perhaps, I shouldn't be all that surprised, considering all the controversy that surrounds evolution. Most schools would probably rather steer clear of the controversy by not teaching evolution.

Now, after my high school days, I didn't take any more biology classes, but I did attend a debate between Duane Gish of the Creation Research Institute and a University of Oregon biology professor. I remember thinking that Dr. Gish sure got the better of his debating partner. Of course that poor old biology professor, he knew his stuff, he just wasn't used to debating a guy who debates for a living. When you decide to debate a Creationist, whether it's Dr. Dino, Duane Gish, or William Dembski, it's probably a no win situation.

So, when did I learn about evolution? Well I probably learned most of what I know from watching the Discovery Channel or visiting museums. You know, I never remember reacting to those displays – so somehow I must have compartmentalized things.

But, with all the polls saying that 50% or more of Americans reject evolution and choose creationism or Intelligent Design instead, then obviously we've not learned much – despite what we see on TV or see in museums. I simply don't believe the reason why we believe this is that as a theory evolution is in trouble or full of holes. Scientists might be reluctant to change at times, but they're not stupid. And as for us religious teachers, we've not done a very good job of helping our congregations learn how to read the Bible critically as well as reverently.

Reading books liked Edward Humes' *Monkey Girl* and *Not in our Classrooms,* (edited by Eugenie Scott and Glenn Branch) help explain why it is we as a nation know so little about the theory of evolution – we simply haven't been taught the theory!

As Edward Humes points out there's a myth that evolution won the day at the first Scopes trial in 1925. Yes, William Jennings Bryan was humiliated but the court upheld the Butler Act, and after that references to Darwin were essentially removed from text books and they didn't reappear until the 1960s.

> The reappearance of evolution in America's textbooks led to the resurgence of the long-dormant anti-evolution crusade, and the conflict soon ended in the courts again (*Monkey Girl,* p. 54).

And we've been fighting ever since!

Ponderings on a Faith Journey (Blog)
April 21, 2007

7

THE EARTH IS THE LORD'S, SO TAKE CARE OF IT!

Whether it's global warming, air pollution, lack of safe drinking water, or the extinction of species, from the looks of things we humans have created a mess. It was for this reason that Earth Day was born in 1970. Inspired by a devastating 1969 oil spill off our own Santa Barbara County coast, a movement was born that called the nation's attention to the fact that we had clogged our rivers and streams and fouled our air with any number of pollutants, making the earth less livable for all of God's creatures. Much progress has been made since then, but work remains to be done.

In recent years the issue of climate change has grabbed our attention. Although some in national leadership pooh-pooh global warming as some kind of environmentalist scam, and some preachers have called this ecological movement a Satanic distraction, the scientific evidence continues to mount that we humans contribute significantly to a burgeoning crisis. If current trends continue, we will likely see increased drought, the melting of the polar ice caps – hastening the extinction of species such as the polar bears and rising sea levels, which would displace millions of people. Deadly storms such as Katrina could become more frequent. So, if there's still time to turn things around, what can we do?

I find the biblical injunction that "the earth is the Lord's" compelling. If the earth belongs to the Lord, what's my responsibility for its welfare? I could begin by listening for God's voice emanating from the earth itself. St. Paul offers the image of the creation "groaning in labor pains" waiting for its redemption (Romans 8:22-23). And, if I understand my faith correctly, God will act redemptively through us, which means we have a divine mandate to care for that which God has given us.

There are a number of statements written from a faith perspective about the environment, but I think this particular statement – "An Evangelical Statement on the Care of Creation" (www.creationcare.org) – catches well the ideal that we're responsible for the environment.

> Because we await the time when even the groaning creation will be restored to wholeness, we commit ourselves to work vigorously to protect and heal that creation for the honor and glory of the Creator . . . We and our children face a growing crisis in the health of the creation in which we are embedded, and through which, by God's grace, we are sustained. Yet we continue to degrade that creation.

In response to this call for repentance, I confess that I'm not as environmentally sensitive as I should be. I waste too much water, gasoline, and electricity. I contribute more than my share of garbage to the landfills. Yes, I've tried to be more responsible: I drive a compact car and have exchanged all but a few light bulbs for fluorescents. I'm fortunate to live in a climate that's neither too hot nor too cold, so my use of heat and air conditioning is limited. But if I lived elsewhere, it might not be quite so easy to be a good steward of these energy resources. While I try to do my part, it's not nearly enough. What is true for me is likely true of Americans in general.

Confession is a start, but Earth Day is a call to action. It calls us to limit our ecological footprint and reclaim the environment. To do so isn't to worship nature, but rather it's recognition of a gift to be treasured and cared for.

How often do we hear that making fuel efficient vehicles or finding alternative sources of energy is too costly? Where, I wonder is the American sense of ingenuity and creativity? Much of our technology is decades old. The incandescent bulb hasn't changed all that much in fifty years, and electricity is transported through the same inefficient lines as when I was a child. Yes, it will take some money and some political will power to make changes, but progress never comes easy or cheaply, at least in the short run. But if we will commit ourselves to protecting and reclaiming the environment, and make the difficult choices now rather than later, then I expect that in the long term we will reap great benefits that can be shared across the globe.

If the earth is the Lord's, then let's celebrate Earth Day by heeding the call to redeem God's creation so that later generations will have an earth to enjoy!

Lompoc Record
Faith in the Public Square
April 22, 2007

8

LISTENING FOR NATURE'S VOICE

Global warming, Al Gore says, is an inconvenient truth. We are seeing patterns of climate change that include changes in migration patterns, a shrinking Arctic ice pack, huge ice sheets breaking free from Antarctica, and the shrinking of glaciers worldwide. Some of this may be due to natural cycles, but there's also clear evidence that human actions are speeding up the process. Besides climate change, we've seen holes appear in the ozone layer, the disappearance of wet lands and rain forests, and an increase in pollution in all its forms. While industrialization has improved life, improvement has come with unfortunate side effects. Finding the right balance won't be easy, and it likely will prove costly, but the alternative is leaving an uninhabitable world to our descendants.

There's a phrase in Paul's letter to the Romans that seems appropriate. He says that the "whole creation has been groaning in labor pains until now" (Romans 8:22). I doubt if Paul had our modern environmental concerns in mind, but perhaps there's something to be said here about listening to nature's groans for God's voice. If so, how might we listen for this voice?

Today is the third annual observance of Evolution Sunday (now Evolution Weekend). Nearly 800 congregations – Jewish, Christian, and Unitarian among others – are observing the birthday of Charles Darwin, a figure in history firmly connected to the theory of evolution and controversial in many religious communities as a symbol of antireligious sentiment. I'm not a scientist and don't pretend to be one. Instead, I'm a pastor, trained in theology, who is concerned about what some consider to be a war between religion and science. This observance of Darwin's birthday, as I see it, is a call for people of faith and people of science to listen to each other for the good of the planet.

As a Christian, I look to the Bible for a word from God, and in doing this I'm no different from people of most religious faiths, who also turn to their sacred scriptures to hear that voice of guidance. While looking first to Scripture, I believe God has other ways of speaking to us. Perhaps God is speaking to us in the groans of the world, inviting us to do something redemptive. If so, then how might we interpret these groans? What tools would we need to make sense of them?

I've come to the conclusion that if we're going to hear this voice, we must turn to science. Unfortunately, many people of faith have become jaded about science. Having been led both by some religious teachers and by some scientists to believe they must choose – the Bible or the Science Book – they choose the Bible and neglect science. Indeed, not only do they neglect it, they revile it. So, when scientists tell us that the earth is warming in ways that are destructive, many of us won't listen. As a result of our inability to hear nature's cry, we continue to overuse, misuse, and destroy the earth.

Seeking a sense of balance, I've developed a new appreciation for the message of Genesis 1. First of all, it tells us that the created order is good. Secondly, it calls for us to be good stewards of this created order. Older translations speak of dominion, a word that has been taken as permission to despoil the earth, but I believe the intent of the passage is different. It is a call to tend to the needs of creation and live in peace with it.

It's the sense of urgency about the state of nature that led me to embrace this move to bridge the gap between faith and science, and become an early signer of the *Clergy Letter Concerning Religion and Science* (being one of more than 11,000 persons to do so). It's also why our congregation is again observing Evolution Weekend.

In the context of this observance, I ask the question: "Can we hear God's voice speaking to us through nature's voice?" If so, then what is God saying? Perhaps it is this: "Be good stewards of my creation." Then, as we seek to be good stewards, perhaps we could start the process by simply replacing an incandescent bulb or two with a CFL bulb. We could decide to drive a smaller car or car pool. If we all do something, then together we can make a difference in the fate of the created order.

Lompoc Record
Faith in the Public Square
February 10, 2008

9

DEFINING SCIENCE

There is this constant argument going on between pro- and anti-evolutionist groups. That argument centers on the definition of science, with both sides charging the other with trading in ideology rather than science. Standing at the center of the debate is the nature of scientific study itself.

I am not a scientist, even though I like reading about science. This means that there is much about science that I don't know. I hadn't thought much about it before reading Karl Giberson's *Saving Darwin* (Harper One, 2008), but I think I finally understand why we keep having such unfruitful discussions about evolution. The problem is that our definition of the scientific method essentially derives from physics, not biology, and when you use a physics-based definition for evolution it doesn't quite work. Biology isn't the same as physics.

Giberson notes that many creationists and Intelligent Design theorists like to point to a definition set out by Karl Popper, the philosopher of science. It's a method modeled by Einstein's theory of relativity. Einstein had set out the principle of relativity, suggesting that light would be warped by space. An astronomer picked up on the challenge, tested it, and proved it to be correct.

Giberson notes that Popper took this situation and based his own definition of science upon it, thus:

> All genuinely scientific theories, he argued, must make novel predictions about unknown phenomena. These predictions must be articulated so clearly that they can be conclusively refuted by observation. And if the predictions fail, the theory has been falsified. If a theory cannot make such falsifiable predictions, then it cannot claim to be scientific (p. 186).

Because evolutionary theory is complex and dependent on history, it didn't lend itself well to Popper's definition. This lead him to declare that Darwinism, along with Marxism and Freudianism, were pseudosciences.

Creationists have picked up on this and charged that evolution is not science but ideology. Now as Giberson points out, Popper later recanted, but Creationists and ID folk don't note that change of heart. Unfortunately, many supporters of evolution make use of this same definition in arguing against Creationism and ID – so both sides are making major mistakes.

> Physics-based theories present their conclusions in tidy mathematical equations – think E=mc2. The relevant phenomena can be demonstrated in laboratory experiments and in public displays at science museums. Impressive technological spin-offs bathe the underlying science in the warm glow of credibility. Evolution, alas, offers nothing but vague generalities – "the fittest survive" – and invokes entities like "common ancestors" or processes like "speciation," for which the evidence is often depressingly small and indirect. . . . (p. 188).

Giberson goes onto note that the disciplines are quite different and so the definitions don't apply equally. Physics is simple and thus words like elegance and beauty fit well. Their theories are neat and testable, but biology is quite different. Life is complex, the processes ever changing, and except for the fossil record, we don't have much to go on.

So, we have a conundrum to consider: How do you define science? It's not as easy to do as we once thought!

Ponderings on a Faith Journey (altered)
February 22, 2009

10

WHY SHOULD PEOPLE OF FAITH
OBSERVE EVOLUTION WEEKEND?

It's been said that science and religion are at war, and the partisans on both sides seem to agree on one thing: If evolution is true, then God doesn't exist.

Because biblical literalists and proponents of "Intelligent Design" have captured the attention of the media and provided fodder for atheists such as Sam Harris and Richard Dawkins, it might appear that there are only two choices: atheism and science or belief and biblical literalism. There is, however, a middle ground, and many of us in the faith community believe that it's unnecessary to make a choice between God and Darwin.

Although I'm concerned about what some call scientific materialism, I'm equally concerned about what I perceive to be an anti-science perspective that goes under the guise of "scientific creationism," or its more respectable version – "Intelligent Design." These efforts have sought to undermine the scientific consensus that holds evolution to be the explanation for the origins and development of life on earth.

These challenges have more to do with religion than science, but unfortunately in their attempt to wrap religious doctrine in scientific language, they have influenced public perception of science and in the course of time have done harm to both to science and to faith. It has undermined the intellectual credibility of our faith traditions, especially the Christian faith, and it could have catastrophic implications for the environment, for medicine, and for our economy.

Consider for a moment this fact – most of our medical advancements, the ones that save lives every day, are predicated on the theory of evolution. In addition, the growing skepticism about science has led to a rejection of the scientific consensus that we are experiencing global warming, and I believe this skepticism that has religious roots, has discouraged young people from pursuing the study of science.

Some of us who are concerned about the current state of affairs have been observing Evolution Weekend (Feb. 10-12 this year). Since 2006, hundreds of congregations from across the country have stepped forward and declared that the planet and its inhabitants require our support. Clergy like me are recognizing that we have a voice that needs to be heard, if for no other reason than that important scientific discoveries could be delayed or dispensed with by religiously motivated opponents to science.

Participants in Evolution Weekend have been accused of participating – perhaps unwittingly – in a grand ruse or conspiracy to introduce bad science and atheistic ideology into our schools. We have been told that the vast majority of Americans reject Darwinism – and thus the reigning theory of evolution – because they see through the science and ideology.

Speaking for myself, I must say, I've not been duped by a Darwinist conspiracy. I have nothing to gain from such participation, except that it arises from a concern for important scientific challenges – such as global warming – and from a concern for the reasonableness of my faith profession. Not being a scientist, I cannot speak to the intricacies of the evolution or Darwin's theory of evolution by natural selection. Yet a common sense look at the issues suggests that Darwin might be right.

As a person of faith who seeks truth, wherever it leads, I'm convinced that while there may be disagreements on the details of the theory of evolution, they are in agreement as to the basic premise that all life has a common origin. In making this affirmation I don't support a secularist agenda, for I do believe in God the Creator, but I believe that science must inform my understanding of God's creative ways. From Darwin's time to the present, good Christian theologians have been able to reconcile the two. It's not just liberals, but even conservatives such as Benjamin B. Warfield have sought to find common ground. Why? It is a commitment to follow facts, as they're made known ,where they lead.

Each participating congregation has in its own way of observing Evolution Weekend. But, we hold in common this desire – to address the fears of those who believe that Darwin and evolutionary theory are a threat to faith. In response to these fears, we suggest that people of faith find a way to build a bridge of understanding. Although this project is relatively new, the need for religious people to address this issue isn't new. In each new generation theologians and religious leaders have stepped forward to deal with the issue.

"Evolution Weekend" grew out of an open letter written by a science professor calling on clergy to voice concern about efforts to undermine the teaching of evolution in our schools. I was one of the early signatories, and in time the "Clergy Letter Project" has garnered more than 10,000 signatures. Among the signatories are well-known theologians and Bible scholars, but the majority are local clergy from a wide range of denominations and faith traditions, from evangelical to Muslim. Those of us who have signed the letter or have participated in this project haven't celebrated either atheism or scientific materialism, but we have sought to hold up the importance of a reasonable faith. Therefore, as a person of faith, I have staked a claim in the debate and have refused to capitulate to those who want us to make either/or choices – God or Darwin.

As a Christian, I affirm God's intimate involvement in the creation of the universe, but I also recognize that such beautiful examples of God's handiwork as Crater Lake and Mount Shasta (I grew up in close proximity to both), the Grand Canyon, the Sleeping Bear Dunes, as well as our human bodies have natural explanations. Both the scientist and the theologian describe the same phenomenon but they use different vocabulary and tools to do so. If we are willing to recognize and affirm the contributions of both the scientist and the theologian there can be a meaningful and profitable conversation – a conversation that has important consequences for human society and the planet we inhabit.

Troy Patch
February 9, 2012

11

Politics, Theology, and the Environment

In recent days politics and theology have become intertwined, with presidential candidates debating the theological veracity of their opponents. The most telling example was Rick Santorum's charge that President Obama has embraced a "phony theology" that isn't in line with what the Bible teaches. Elsewhere I addressed this charge, noting that to say that something is phony is to raise questions of a person's religious integrity.

When the "charge" was first made, the former Senator didn't elaborate, but a few days later he backed off a bit and spoke of the President's supposedly radical environmentalist world view that he believes is rooted in a theology (world view) that lacks biblical support. In this clarifying statement about what he meant by a "phony theology" that lacks biblical warrant, Santorum made the following statement:

> "That's why I was talking about energy. This idea that man is here to serve the earth, as opposed to husband its resources and being good stewards of the earth, and I think that is a phony ideal."

Many people believe that religion is private and should remain private. I believe that religion is personal, but that it also has public implications.

What we believe about God impacts the way we see the world. In this, I'm in agreement with Sen. Santorum, though I strongly disagree with the way he understands both the Bible and Christian theology. I will also admit that my politics is closer to that of the President than that of the former Senator's. But for a moment I'd like to have us put aside partisan politics and consider the theology of creation (I use creation here in a theological sense, not a scientific one).

I believe that a good case can be made that concern for the environment is deeply rooted in Scripture and Christian theology, and that it is a moral imperative for us to concern ourselves with protecting and preserving the environment.

If the earth exists for the benefit of humanity, does that mean that it exists solely for our benefit? Does it mean that we have the right and obligation to "husband its resources" without concerning ourselves with the long term viability of nature?

At the core of this debate is the definition of stewardship. Does being a good steward mean taking good care of something entrusted to us, or does it imply use of something for our immediate benefit? In answering these questions I would like to suggest that Sen. Santorum's views, which he claims are biblical, could be out of line not only with Scripture, but also the teaching of the Roman Catholic Church, of which he is a member.

Before I turn to Catholic teaching on this issue, I'd like to suggest an alternative reading of the biblical story concerning creation and our responsibility to it. In Genesis 1 we find the first creation story. Throughout this beautiful poetic statement, we hear God say of nature – "It is good." As God concludes the work of creation God creates humankind in God's image, male and female, and entrusts to them (us) the responsibility of being stewards of creation. Stewardship doesn't entail absolute rule. There is always accountability to God, who created humankind to represent God in creation. We are not above creation, we are part of it. Being that we're part of creation, we have an important role to respect and preserve the world God has created, and use its resources appropriately.

To add to this point about our place in creation, consider the message found in the designated reading from the Hebrew Bible for the First Sunday of Lent (this Sunday). The text is Genesis 9:8-17. It's part of the Flood story, and as Noah, his family, and the animals disembark from the ark, God makes a covenant, not just with Noah, his family, and his descendants, but also "with every living being with you – with the birds, with the large animals, and with all the animals of the earth, leaving the ark with you. I will set up my covenant with you so that never again will all life be cut off by floodwaters. There will never again be a flood to destroy the earth" (Gen. 9:9-11 CEB). God places in the heavens a bow, a rainbow, to serve as a reminder to God of this covenant, which God makes with "all creatures on earth." Humanity is, in this scenario, part of, but not the sole concern of God's covenant with the earth.

I take these texts as an encouragement to be good stewards of the environment. Thus, I believe a truly biblical theology will recognize that our welfare as human beings is linked to that of the rest of creation. My well-being is affected by the despoiling of the environment, whether that is pollution or overuse of natural resources. Although the Senator suggests that global warming is not science but politics, I beg to differ. If climate science is correct, we could be setting ourselves on an environmentally unsustainable course that could be as destructive to the future well-being of humanity as any other economic consideration.

I've mentioned but two biblical passages that, in my view, call for us to attend to the protection of the environment. I would suggest that this is part of our covenant responsibility. Yes, we do have a special role as God's representatives in the world, but that responsibility is an ecologically sustainable one. We can make use of earth's resources, but we should do this in a wise and sustainable manner.

I said earlier that a biblical case can be made for what is often called "creation care," but it is also a central focus of Roman Catholic Social Teaching. Although I don't agree with the Catholic Church on every issue, there are many places where I am in agreement with the teachings of this church. I would say that on the issue of the environment, I'm probably closer to the Church's teachings than is Sen. Santorum. In support of my claim I want to point out a statement from the United States Conference of Catholic Bishops and Pope John Paul II.

First consider this statement from the USCCB on Caring for God's Creation

> 54. We show our respect for the Creator by our steward-
> ship of God's creation. Care for the earth is a duty of our faith
> and a sign of our concern for all people. We should strive to live
> simply to meet the needs of the present without compromising
> the ability of future generations to meet their own needs. We
> have a moral obligation to protect the planet on which we live—
> to respect God's creation and to ensure a safe and hospitable
> environment for human beings, especially children at their
> most vulnerable stages of development. As stewards called by
> God to share the responsibility for the future of the earth, we
> should work for a world in which people respect and protect
> all of creation and seek to live simply in harmony with it for
> the sake of future generations.

Pope John Paul II is even clearer and stronger in his statements about our responsibilities toward the environment. These are stated powerfully in a 1990 address celebrating the World Day of Peace. I'd like to share just the two opening paragraphs of this statement, and invite you to read the entire message in support of this statement. The late Catholic leader declared:

> In our day, there is a growing awareness that world peace
> is threatened not only by the arms race, regional conflicts and
> continued injustices among peoples and nations, but also by
> a lack of *due respect for nature*, by the plundering of natural
> resources and by a progressive decline in the quality of life.
> The sense of precariousness and insecurity that such a situation
> engenders is a seedbed for collective selfishness, disregard for
> others and dishonesty.

Faced with the widespread destruction of the environment, people everywhere are coming to understand that we cannot continue to use the goods of the earth as we have in the past. The public in general as well as political leaders are concerned about this problem, and experts from a wide range of disciplines are studying its causes. Moreover, a new *ecological awareness* is beginning to emerge which, rather than being downplayed, ought to be encouraged to develop into concrete programmes and initiatives.

Pope John Paul II makes the claim that our relationship to the environment is a moral issue. He states that the future sustainability of this world depends on our ability to address this issue, including the question of climate change. He also notes that how the developed world uses resources has implications for developing nations and for those persons living in poverty. Thus, it is a matter of addressing the needs of "the least of these" (Matthew 25).

I recognize there are differences of opinion on matters theological and political when it comes to the environment. But, a commitment to protecting the earth and its resources – a concern for the welfare of creation as a whole – is not foreign to the biblical story or to mainstream Christian theology. In fact, as I read scripture and theology, I would suggest that it is a divine imperative that we concern ourselves with protecting the environment as an expression of our calling to be God's stewards of creation. It is also, in our best interests to concern ourselves with this calling.

Troy Patch
February 26, 2012

12

Beware the God of the Gaps

It's said by some that we live in a de-enchanted world. From at least the days of Isaac Newton, science has peeled back the layers of reality and offered explanations for what was once believed to be the realm and work of God. What was once unexplainable now has a natural or rational explanation. The universe is governed by the laws of nature, ever evolving and developing without the apparent need for divine intervention. Newton did believe that from time to time God was needed to fine tune things, but in general, things ran rather mechanically. God might be the first cause, but now has been relegated to a distant, behind the scenes, role in the universe. If God is so distant and uninvolved, then surely we can ignore God – and we often do – even if we profess a particular faith.

The question that modern science raises for people of faith is a difficult one to answer. Whereas once many diseases or psychological conditions had spiritual explanations (divine punishment or demon possession), now we can explain them in medical terms and find solutions devised by science. Where we once believed that the earth stood at the center of the universe, with the sun, the planets, and the stars rotating around us, we now know we're nothing but the third rock rotating around a rather small and insignificant star.

So, is there a place for God in this new age of science? There are some who answer the question by giving priority to religious explanations and downplaying science, but is this wise?

Consider for a moment the debate about climate change. Why are so many people resistant to the idea that not only are humans contributing to global warming, but that there is consensus within the scientific community concerning this theory. Why do so many people treat scientific "theories" as if they are mere conjecture or opinion? Is there a fear that if we give room to science, we might lose God? Have we bought into the premise offered by folks such as Richard Dawkins that theology is nothing more than "fairiology?" In other words, theology isn't a real intellectual pursuit – it's simply superstition.

As we approach the birthday of Charles Darwin (February 12) I believe it is important that people of faith address the question of whether we can continue believing in God in an age of science. If Darwin is correct, and I believe he is, that we have evolved from a common ancestor over a period of billions of years, then where does God fit? Some will answer that we must adhere to the biblical story, but is the biblical story designed to tell us scientific truth? Is that its purpose? Some will say that evolution has dangerous consequences, so it needs to be resisted. After all, didn't Hitler believe in evolution? That people have twisted theories for their own benefit doesn't mean that they're not true.

If science is correct on matters such as evolution, the question facing us is this – where does God fit? In answer to the question I've posed a warning: "Beware the God of the Gaps." While I will affirm the premise that God is the Creator of the Heavens and the Earth, I am concerned about the tendency among some of my co-religionists to use God as an explanation for the unexplainable. We call this the "God of the Gaps," using God as a stopgap measure to explain the unexplainable. So, because we don't know exactly how things began, except that it seems to have started with a "Big Bang," then surely we can say that God is the "First Cause." Once things got started, a further issue arises – how did life forms develop? Proponents of Intelligent Design suggest that God is responsible for the design of the forms that life takes, using the idea of "irreducible complexity" as proof that a designer is needed. As William Paley, an 18th century Anglican priest suggested, if you find a watch lying alongside the road, you must assume it had a creator – a watchmaker. Since nature seems to express an intelligent design, then we can assume it has a designer – a divine watchmaker. And with this assumption, we have proof for the existence of God.

The problem with Paley's argument is that the realm of God's activity shrinks with every scientific discovery. What we can't explain today, could receive an explanation tomorrow. Theologian Dietrich Bonhoeffer worried about this tendency, and in a letter written from prison to his friend and Eberhard Bethge, Bonhoeffer writes that we shouldn't use God as a "stopgap for the incompleteness of our knowledge, because then, as is objectively inevitable–when the boundaries of knowledge are pushed ever further, God too is pushed further away and thus is ever on the retreat."[1] Bonhoeffer wisely recommends that we seek to find God present in what we know, rather than what we don't know. Rather than thinking of God as the divine watchmaker, we must assume that God is present in and with the development of this universe.

I'm not a scientist; I'm trained as a historian and a theologian. As a Christian who is by profession a pastor of a church, I believe in God. By faith I affirm that God is the Creator of the Heavens and the Earth. Having said this, I don't expect the Bible, to which I turn for my spiritual sustenance, to offer a scientific account of reality. That's not the purpose of the Bible. To ask it to offer such explanations is to force it to do something it's not "designed" to do. For such explanations, I turn to science. I don't need to force scripture and science to say the same thing. By faith, I can, however, affirm the premise that the heavens and earth, whatever the scientific explanations might be, do declare the Glory of God.

Because I am increasingly concerned about the implications for both the faith community and the world at large, I have invested myself in building bridges between the scientific and the religious. I believe there needs to be conversation, with both realms of thought learning from the other. With this in mind I have been a participant in the Evolution Weekend project from its very beginning. This weekend congregations from a variety of faith traditions from across the nation and beyond will observe this event as a way of saying no to those who would appeal to a stopgap God and to

---

1 Dietrich Bonhoeffer, *Letters and Papers from Prison. Dietrich Bonhoeffer Works,* (Fortress Press, 2009), pp. 405-406

those who reject outright any place for God in the conversation. Our future as a planet depends in large part in the success of this conversation. So, won't you join me in this observance? Let us put an end to the "war" between religion and science so that we can live together in a more peaceful, just, and fruitful world. For me, that means that God is ever present in this process, not sitting back, having finished everything on the sixth day!

Troy Patch
February 8, 2013

Appendix

Liturgical Resources

Calls to Worship:

L: The Lord Said: "Let there be Light and there was Light"
P: And "It is Good"
L: The Lord Said: "Let there be a dome in the midst of the waters:
P: And "It is Good"
L: The Lord Said: "Let the waters be gathered so the dry land can appear"
P: And "It is Good."
L: The Lord Said: "Let there be vegetation on the land"
P: And "It is good."
L: The Lord said: "Let there be lights in the sky"
P: And "It is Good."
L: The Lord said: "Let the heavens, the earth, and the waters team with life"
P: And: "It is Good."
L: The Lord said: "Let's make humanity in our image, and invite humanity to care for the earth"
P: And: "It is Good."
L: The Lord said: "Now it's time to rest"
P: And: "That's Good also."

———

L: The heavens are telling the glory of God.
P: And the firmament proclaims his handiwork.

L: In the heavens he has set a tent for the sun,

P: Its rising is from the end of the heavens, and its circuit to the end of them; nothing is hide from its heat.

L: Let the words of my mouth and the meditation of my heart be acceptable to you, O Lord, my Rock and my Redeemer. (Based on Psalm 19)

———

L: Open your eyes and take in the sights of God's Creation.

P: What beauty lies before us.

L: Open your ears and listen for the sounds Creation brings forth

P: What glories, these sounds declare.

L: Open your mouths and declare the Glory of the Creator

P: We give thanks for the gift of Creation, for with it comes life in all its dimensions.

PRAYERS

Invocations

Holy One, Creator of the universe, you have filled your creation with surpassing opportunities and potential for growth. You have given us the opportunity to live within this creation, with an abundance of gifts to share in. As we give thanks for your gifts, we pray that we would be careful stewards of these gifts. May they inspire us to worship you in all your glory, glory that you share with us each day. May our hearts and minds draw near, and our voices ring out with praise, for you are the blessed Creator. Amen.

Almighty and everlasting God, you made the universe with all its marvelous order, its atoms, worlds, and galaxies, and the infinite complexity of living creatures: grant that as we probe the mysteries of your creation, we may come to know you more truly, and surely fulfill our role in your eternal purpose; in the name of Jesus Christ our Lord. Amen.[1]

Stewardship Prayer

God of abundance and grace, you have blessed us with many wondrous gifts. In the beginning of all things, you blessed an expansive universe with the potential to give birth to life in wondrous variety. And to humanity, whom you created in your image, you gave the responsibility for stewardship, to take care of the garden, that it might bring forth your abundance. As we come offering these signs of gratitude, may we also take to heart your calling to be careful stewards of your creation. Amen.

Communion Prayers

Bread and wine, these are your gifts to us. They remind us of Jesus' body and blood, given that we might find life in you. We remember as we take these symbols into our bodies, that your word took on bodily form, and in doing so you affirmed the blessings of human life. You declared your love for this world, the world of your creation. We ask that you bless these elements which have their origin in the earth you have created so that we might enjoy life in all its fullness. Amen.

We have gathered at your table. We have shared in the bounty of your creation. We have remembered the one who embodied your love for the world. Having fed upon this feast and having been nourished by your presence in this feast, may we go forth from this table and look for your presence in the created order. May we go forth giving thanks for all that we enjoy in life. May we give thanks for the scientists and the theologians who have taught us to

1 *The Book of Common Prayer: According to the use of the Episcopal Church*, (1979), p. 827.

value this good earth, so that we might take good care of that which emerged over time, but also with your loving embrace. Amen.

ALSO FROM ENERGION PUBLICATIONS
ON CREATION

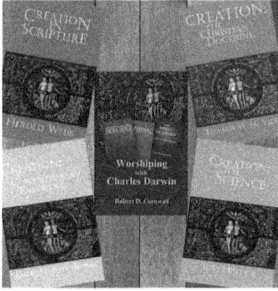

These five books take a long at creation as understood in the Christian faith from the perspective of different disciplines. In *Creation in Scripture*, the perspective is biblical studies. In *Creation: The Christian Doctrine*, it is systematic theology. In *Creation: The Science* it is, of course, science. In *Creation in Contemporary Experience*, we see creation from the perspective of our ongoing experience, ethics, and sociology. And of course in this volume we have taken a look from the perspective of a pastor, looking at worship, and liturgy.

Creation in Scripture	Herold Weiss	$12.99
Creation: the Christian Doctrine	Edward W. H. Vick	$12.99

Forthcoming:

Creation: the Science	Tony Mitchell	$12.99
Creation in Contemporary Experience	David Moffett-Moore	$12.99

ALSO BY THE AUTHOR

Both accessible and insightful, *Faith in the Public Square* is a must read for anyone seeking greater understanding of societal trends in 21st century America.

Saeed A. Khan
Lecturer, Wayne State University
Fellow, Institute of Social Policy and Understanding

FAITH
in the
Public Square

LIVING FAITHFULLY
IN
21ST CENTURY AMERICA

Robert D. Cornwall

MORE FROM ENERGION PUBLICATIONS

Personal Study

Finding My Way in Christianity	Herold Weiss	$16.99
Holy Smoke! Unholy Fire	Bob McKibben	$14.99
The Jesus Paradigm	David Alan Black	$17.99
When People Speak for God	Henry Neufeld	$17.99
The Sacred Journey	Chris Surber	$11.99

Christian Living

Faith in the Public Square	Robert D. Cornwall	$16.99
Grief: Finding the Candle of Light	Jody Neufeld	$8.99
Crossing the Street	Robert LaRochelle	$16.99

Bible Study

Learning and Living Scripture	Lentz/Neufeld	$12.99
From Inspiration to Understanding	Edward W. H. Vick	$24.99
Luke: A Participatory Study Guide	Geoffrey Lentz	$8.99
Philippians: A Participatory Study Guide	Bruce Epperly	$9.99
Ephesians: A Participatory Study Guide	Robert D. Cornwall	$9.99

Theology

Creation in Scripture	Herold Weiss	$12.99
Creation: the Christian Doctrine	Edward W. H. Vick	$12.99
The Politics of Witness	Allan R. Bevere	$9.99
Ultimate Allegiance	Robert D. Cornwall	$9.99
History and Christian Faith	Edward W. H. Vick	$9.99
The Church Under the Cross	William Powell Tuck	$11.99
The Journey to the Undiscovered Country	William Powell Tuck	$9.99
Eschatology: A Participatory Study Guide	Edward W. H. Vick	$9.99

Ministry

Clergy Table Talk	Kent Ira Groff	$9.99
Out of This World	Darren McClellan	$24.99

Generous Quantity Discounts Available
Dealer Inquiries Welcome
Energion Publications — P.O. Box 841
Gonzalez, FL_ 32560
Website: http://energionpubs.com
Phone: (850) 525-3916

www.ingramcontent.com/pod-product-compliance
Lightning Source LLC
Chambersburg PA
CBHW031558040426
42452CB00006B/344